# RETHINKING MATHEMATICAL CONCEPTS

# ELLIS HORWOOD SERIES IN
# MATHEMATICS AND ITS APPLICATIONS

*Series Editor:* Professor G. M. BELL, Chelsea College, University of London

The works in this series will survey recent research, and introduce new areas and up-to-date mathematical methods. Undergraduate texts on established topics will stimulate student interest by including present-day applications, and the series can also include selected volumes of lecture notes on important topics which need quick and early publication.

In all three ways it is hoped to render a valuable service to those who learn, teach, develop and use mathematics.

**MATHEMATICAL THEORY OF WAVE MOTION**
G. R. BALDOCK and T. BRIDGEMAN, University of Liverpool.
**MATHEMATICAL MODELS IN SOCIAL MANAGEMENT AND LIFE SCIENCES**
D. N. BURGHES and A. D. WOOD, Cranfield Institute of Technology.
**MODERN INTRODUCTION TO CLASSICAL MECHANICS AND CONTROL**
D. N. BURGHES, Cranfield Institute of Technology and A. DOWNS, Sheffield University.
**CONTROL AND OPTIMAL CONTROL**
D. N. BURGHES, Cranfield Institute of Technology and A. GRAHAM, The Open University, Milton Keynes.
**TEXTBOOK OF DYNAMICS**
F. CHORLTON, University of Aston, Birmingham.
**VECTOR AND TENSOR METHODS**
F. CHORLTON, University of Aston, Birmingham.
**TECHNIQUES IN OPERATIONAL RESEARCH**
**VOLUME 1: QUEUEING SYSTEMS**
**VOLUME 2: MODELS, SEARCH, RANDOMIZATION**
B. CONNOLLY, Chelsea College, University of London
**MATHEMATICS FOR THE BIOSCIENCES**
G. EASON, C. W. COLES, G. GETTINBY, University of Strathclyde.
**HANDBOOK OF HYPERGEOMETRIC INTEGRALS: Theory, Applications, Tables, Computer Programs**
H. EXTON, The Polytechnic, Preston.
**MULTIPLE HYPERGEOMETRIC FUNCTIONS**
H. EXTON, The Polytechnic, Preston
**COMPUTATIONAL GEOMETRY FOR DESIGN AND MANUFACTURE**
I. D. FAUX and M. J. PRATT, Cranfield Institute of Technology.
**APPLIED LINEAR ALGEBRA**
R. J. GOULT, Cranfield Institute of Technology.
**MATRIX THEORY AND APPLICATIONS FOR ENGINEERS AND MATHEMATICIANS**
A. GRAHAM, The Open University, Milton Keynes.
**APPLIED FUNCTIONAL ANALYSIS**
D. H. GRIFFEL, University of Bristol.
**GENERALISED FUNCTIONS: Theory, Applications**
R. F. HOSKINS, Cranfield Institute of Technology.
**MECHANICS OF CONTINUOUS MEDIA**
S. C. HUNTER, University of Sheffield.
**GAME THEORY: Mathematical Models of Conflict**
A. J. JONES, Royal Holloway College, University of London.
**USING COMPUTERS**
B. L. MEEK and S. FAIRTHORNE, Queen Elizabeth College, University of London.
**SPECTRAL THEORY OF ORDINARY DIFFERENTIAL OPERATORS**
E. MULLER-PFEIFFER, Technical High School, Ergurt.
**SIMULATION CONCEPTS IN MATHEMATICAL MODELLING**
F. OLIVEIRA-PINTO, Chelsea College, University of London.
**ENVIRONMENTAL AERODYNAMICS**
R. S. SCORER, Imperial College of Science and Technology, University of London.
**APPLIED STATISTICAL TECHNIQUES**
K. D. C. STOODLEY, T. LEWIS and C. L. S. STAINTON, University of Bradford.
**LIQUIDS AND THEIR PROPERTIES: A Molecular and Macroscopic Treatise with Applications**
H. N. V. TEMPERLEY, University College of Swansea, University of Wales and D. H. TREVENA, University of Wales, Aberystwyth.
**GRAPH THEORY AND APPLICATIONS**
H. N. V. TEMPERLEY, University College of Swansea.

# RETHINKING
# MATHEMATICAL
# CONCEPTS

ROGER F. WHEELER
Department of Mathematics
University of Leicester

**ELLIS HORWOOD LIMITED**
Publishers · Chichester

Halsted Press: a division of
**JOHN WILEY & SONS**
New York · Chichester · Brisbane · Toronto

First published in 1981 by

**ELLIS HORWOOD LIMITED**

Market Cross House, Cooper Street, Chichester, West Sussex, PO19 1EB, England

*The publisher's colophon is reproduced from James Gillison's drawing of the ancient Market Cross, Chichester.*

**Distributors:**

*Australia, New Zealand, South-east Asia:*
Jacaranda-Wiley Ltd., Jacaranda Press,
JOHN WILEY & SONS INC.,
G.P.O. Box 859, Brisbane, Queensland 40001, Australia

*Canada:*
JOHN WILEY & SONS CANADA LIMITED
22 Worcester Road, Rexdale, Ontario, Canada.

*Europe, Africa:*
JOHN WILEY & SONS LIMITED
Baffins Lane, Chichester, West Sussex, England.

*North and South America and the rest of the world:*
Halsted Press: a division of
JOHN WILEY & SONS
605 Third Avenue, New York, N.Y. 10016, U.S.A.

**British Library Cataloguing in Publication Data**
Wheeler, Roger F.
    Rethinking mathematical concepts.—(Ellis
    Horwood series in mathematics and its
    applications).
    1. Mathematics—1961—
    I. Title
    510      QA36     80-42100
    ISBN 0-85312-284-9 (Ellis Horwood Limited, Publishers)
    ISBN 0-470-27116-7 (Halsted Press)

Typeset by Preface Ltd, Salisbury, Wilts.
and Printed by R. J. Acford Ltd, Chichester.

ISBN 0 85312 308 X
( llis Horwood /

# Table of Contents

# Introduction

This book is addressed to the student mathematics teacher. Its aim is to encourage you to think about the subject and its presentation.

The book may also be of interest to other students in higher education and to all those called on to teach mathematics, especially if this subject has not been your main area of specialization. Indeed, any educator who is keen for students, at whatever level, to acquire a sound grasp of mathematical concepts may find something of novelty in the discussions.

The process of learning mathematics is one of evolution. No topic is ever finally exhausted: as one's experience grows, one continually discovers thrilling new levels at which an idea can be investigated. Ultimately, the challenge of having to teach a topic to someone else reveals a further hierarchy of unsuspected subtleties in the concept every time it is presented. It is important to avoid teaching anything that needs contradicting later: what you say may need further amplification and expansion; hidden assumptions may need to be exposed and examined; the more mature student may demand a more elaborate or precise enunciation. But a wise teacher aims never to have to tell his pupils to forget what they were told last year because he is now going to use a conflicting approach. Textbooks also must be chosen carefully if the teacher is not going to have to say: 'Ignore what the book says: that's not a good way of doing it: I want you to look at it this way'.

This book examines some of the many problems that arise when our subject is presented to pupils. It does not set out to teach you any new mathematics—each chapter assumes you are already acquainted with the essentials of the topic being discussed—and what it certainly does *not* try to do is to develop any theories of mathematics learning or concept formation.

In any sequence of lessons, the teacher has to make many decisions on matters of exposition, particularly of definition and notation, some of which have far-reaching consequences for later work, not always appreciated at the time. As any experienced teacher will tell you, one of the most difficult things in teaching is to get pupils to unlearn something they have been taught; to discard a cherished habit once acquired. If you doubt the truth of this maxim, consider the case of the student who evaluates $\int_0^1 dx/(1 + x^2)$ as $45°$ (and who probably feels aggrieved when you mark it wrong). Having first met the tangent in trigonometry as a function of an angle and grown accustomed to this, he has obviously found it too traumatic to abandon this established mode of thought and to accept that the tangent has long since become a function of a *number*: that $\tan^{-1} 1$ is the number $\frac{1}{4}\pi$, and not an angle. Perhaps this particular hurdle is unavoidable, but many others are not and the fewer obstacles we create for our pupils, the better.

Mathematicians like to feel they are consistent people, but the inherent consistency of their subject does not always carry over to their methods of exposition. One trivial example will suffice for the moment: plenty of more significant examples are discussed later in the book. Ask yourself how you read aloud (as if to a class) the expression $(2x + 3)(x + 5)$. Decide precisely what you actually say before reading on. Did you say '$2x + 3$ into $x + 5$' or '$2x + 3$ times $x + 5$'? You have only to listen to mathematics lecturers and teachers to find out that most people have been brought up to use the reading 'into'. Does this not strike you as an extraordinary reading? Would you use it yourself when teaching a pupil just starting algebra? One hopes not, yet that is just what many teachers are doing. Having spent time in arithmetic lessons clearly associating in the child's mind the word 'into' with the operation of division ('4 into 12 goes 3' and that sort of thing), all of a sudden the teacher expects him in algebra to associate 'into' with the inverse operation of multiplication. The stupidity of this is obvious, yet the conventional presentation of mathematics is shot through with examples like this. One sometimes feels that the surprising thing is not that some children abandon mathematics but that any persevere at all.

During your period of professional training, you will be mainly concerned with trying to evolve exciting ways of presenting your mathematics to your future pupils. But you will also have the opportunity and leisure to reflect on the subject itself. This book will try to give you the background to just a few of the decisions you will have to make once you get to the chalk-face (particularly if you teach pure mathematics at the upper end of a secondary school) and to show you the consequences these have for the development of the subject. The one unforgivable sin in mathematics teaching is to go on following bad models without having any definite reason for doing so. The author dare not expect that you will agree with everything he has written, although he naturally hopes that most of his arguments will be sufficiently cogent to compel assent. If, however, you are not convinced by a particular piece of reasoning and decide to reject the evidence offered, the writer will not be unduly disappointed. At least you will then be acting because of a positive conviction you have arrived at that what you have *chosen* to do provides the best possible approach, and not just continuing to do something in a particular way because 'that's the way I was taught myself' or 'that's the way this textbook does it'. Indeed, one of the things one would most like you to develop is a certain critical faculty in appraising mathematics books (including this one). As a newcomer to teaching, you should certainly *start* by recognizing that the textbook writer will have been more experienced than you and may have had good reasons for the presentation he has chosen but, if I share with you the secret that many books in circulation are seriously misleading in their mathematics, whatever their pedagogic felicities, you will perhaps not feel the need slavishly to follow their authors if you genuinely think you have sound reasons for objecting to their treatment.

This book will try not just to tell you what, in the opinion of the author, the definition of, say, a function should be, but to consider what the alternative definitions are, why the books offer such a bewildering variety, why some are demonstrably unsuitable and, among those that are at least acceptable, what the relative merits are and the reasons why mathematicians are likely to prefer some definitions to others. In other words, it attempts to fill in some of the explanations of *why* certain things are done, whereas books aimed at exposition only tell you *what* is done and *how* it is done. It also considers, for example, questions such as why mathematicians choose *not* to write $\infty = 1/0$: it does not tell you what to say to your pupil when he or she does this; it tries to give you the background you need to have, so that you yourself can devise an explanation that will be appropriate to present to the child. Nor does this book give you any guidance on how to react when you find that your school's chosen examination board has elected to promote the 'arcsin' notation—as, at the time of writing, one board sadly does—if you have become convinced by what you have read that this is manifestly inferior to 'sin$^{-1}$'!

There is one particular trap into which a new teacher can easily fall. He is lured into it because his recollection of the fads and foibles of his mathematics lecturers in tertiary education are necessarily much fresher in his mind than are his memories of the practices of his teachers at school. He is, therefore, strongly tempted to show off his own newly acquired mathematical maturity and imagined sophistication by repeating with his own classes the mannerisms he has picked up since leaving school. But this *may* not be at all desirable. All such fashions need to be inspected closely and their suitability for school use assessed before they are copied with young pupils: this, indeed, is a frequent theme throughout this book.

The point being made is that a lecturer presenting mathematics to a 19 or 20 year old student often expresses himself in a way that he would not (or, at least, should not) if teaching a 15 or 16 year old. As an example, the ways in which he uses the word 'infinity' or the sign '$\infty$' may be particularly lax, because he believes—one hopes the belief is justified!—that he can rely on his listeners' wisdom to translate his sloppy shorthand into taut mathematics. But it will probably be quite inappropriate to copy such imprecise abbreviated statements with schoolchildren, who will not have had a similar depth of experience.

It must be stressed that not all lecturers would subscribe to the above philosophy: this writer, for one, *certainly* would not. If it is wrong to use language or notation in a cavalier way with a pupil, it is no less wrong to do so with an undergraduate. But it is no good shutting one's ears and eyes to the liberties that are taken in practice—one hears and reads them at every turn. It is unrealistic, therefore, to ignore the influence these habits are likely to have been exerting on a candidate new to teaching.

It is hoped that sufficient reminders of the relevant mathematics have been included to make each discussion self-contained, especially where it has been necessary to range beyond the school syllabus in order to support the argument. But since an ordered, logical presentation was neither appropriate nor intended, it may be that some readers will need to refer to explanatory texts to remind themselves of some facts the author has taken for granted. To make the book as useful as possible as a vade-mecum, generous cross-references, both forward and backward, have been provided.

Even if you are a recently qualified teacher of secondary mathematics, you may find the discussions useful and thought-provoking. Indeed, the very experience you have been acquiring may give them a sharper focus, as you may now appreciate, even more than you would have done as a student, the relevance of some of the problems raised. Dare one hope that a few textbook writers and examiners might also find that reading and evaluating the arguments was not a complete waste of time?

Any experienced teacher who dips into this book will recognize the author's debt to the various Mathematical Association Reports, which were a seminal influence in his early years as a schoolmaster and can be warmly recommended to every new teacher. The further debt to my own gifted schoolmasters is one that will not be obvious to the reader, but is just as real.

# Chapter 1

# Relations and operations

---

## 1.1

There are probably three main arguments that would be put forward to justify the introduction of the various changes in the school pure mathematics syllabus during recent decades: (1) that new topics have widened the scope of the curriculum by, for example, applying algebra to interesting structures other than the usual number systems—structures governed by novel rules of operation, such as sets, vectors, matrices, finite arithmetics, permutations and geometric transformations—so that the pupil gets a broader view of the nature of mathematics and its potential areas of application; (2) that there has been an attempt to increase the understanding of mathematical concepts, even at the expense of some loss of proficiency in certain techniques of limited value; and (3) that the idea of mathematics as a language has been promoted by trying to encourage clarity of thought and precision of expression.

In the furtherance of these last two objectives there has been a quite dramatic sharpening of the presentation of fundamental ideas on, for example, operations and relations and functions. You have only to look at the woolly accounts of these notions in some old school textbooks to see what advances in lucidity have been achieved. There are, however, many points worth bringing to your attention before you try your hand at discussing such topics with your classes and the first two chapters are, therefore, devoted to these basic concepts. These chapters may require rather more concentration on your part than some of the later topics in the book. But remember that the more fundamental an idea is for the development of mathematics, the more important it is to make sure that one's own grasp of the concept is secure, so that one's teaching, even with the youngest pupils, will be laying sound foundations.

Not so very long ago, the occurrence in students' work of phrases like

| | | |
|---|---|---|
| '$x$ is $=$ to $y$' | '$x$ and $y$ are $=$' | 'the $=$ numbers $x$ and $y$' |
| 'AB is $\parallel$ to CD' | 'triangles ABC, DEF are $\equiv$' | 'the $\perp$ lines $l, m$' |

was quite common (although it is only fair to point out that perceptive teachers have *always* discouraged any such misuse of mathematical language). But, even if these grammatical mistakes are becoming rarer, they are by no means extinct and, until they are, one cannot be entirely happy about the teaching of relations. One's disquiet is reinforced by the number of student teachers who have been known to say, for example, that the binary relation $\rho$ is symmetric 'if it satisfies $a \, \rho \, b = b \, \rho \, a$'. Please pause for a moment to reflect on the enormity of this blunder!

## 1.2

It will be a good idea to start by looking at some of the (binary) relations that are met in fairly elementary mathematics, together with the symbols conventionally used to represent them, in order to remind ourselves that in all cases the symbol has a *verbal*, not an adjectival, function in a mathematical sentence. (In teaching the subject, of course, valuable use can be made in the early stages of more colloquial relations, such as 'is a brother of', 'lives in the same street as', 'is taught by', and so on.)

$=$ 'is equal to'
$<$ 'is less than'
$>$ 'is greater than'
$\leq$ 'is less than or equal to'
$\geq$ 'is greater than or equal to'
$\mid$ 'divides' [or, less shortly, 'is a divisor of']
$\equiv$ 'is congruent to' [in number theory or in geometry]
R 'is a quadratic residue of'
N 'is a quadratic non-residue of'
$\parallel$ 'is parallel to'
$\perp$ 'is perpendicular to'
$\in$ 'belongs to' [or, less shortly, 'is an element of']
$\subseteq$ 'is a subset of', 'is contained in', or 'is included in'†
$\subset$ 'is a proper subset of', or 'is strictly (properly) contained in'
$\supseteq$ 'contains', or 'includes'

---

†This choice of symbol ($\subseteq$) is *greatly* to be preferred to the extremely unfortunate selection of '$\subset$' to mean 'is a subset of'. Not only does that leave no symbol available for the relation 'is a proper subset of' but the analogy is lost between the partial order relations $\leq$ and $\subseteq$ (which are reflexive, transitive and antisymmetric) and between the dominance relations $<$ and $\subset$ (which are transitive and irreflexive).

⊃ 'strictly (properly) contains'
⇒ 'implies'
⇐ 'follows from' [this reading is shorter than 'is implied by']
⇔ 'implies and follows from'

These last three relations (⇒, ⇐, ⇔) raise such interesting questions that they have been allotted a chapter to themselves [Chapter 17].

Associated with each of these relations is a *complementary* relation, which holds whenever the original relation does not; thus, ≠ ('is not equal to'), ∤ ('does not divide'), ≢, ∉, and so on. The phrases defining a pair of complementary relations are mutual negations. Obviously, R and N are complementary and ≮ is equivalent to ≥; but, of course, ⊄ is not equivalent to ⊇! In practice, some of the relation *symbols* are seldom negated.

All these mathematical relations have very precise meanings and are sharply defined within their contexts, so that no liberties are possible with any of the signs (although the tense or mood of the verb may change in subordinate clauses, as explained in Section 1.7). It is their very precision that makes them a vital part of the language of mathematics: and the less *that* language is abused the better. Sometimes lazy pupils try to get the best of both worlds by obligingly using ⊥ for 'is perpendicular to', but inventing a symbol like ⊥$^r$ or ⊥$^{ar}$ for the adjective or noun 'perpendicular'. This is still thoroughly bad practice, because it debases the currency of the relation symbol '⊥'.

## 1.3

The long-suffering sign for the relation of equality always has been (and probably always will be) subject to the worst mangling, with pupils using it as an all-purpose link to mean 'I think there is some connexion, possibly rather tenuous, between what I have written just before and just after this sign, which I hope you may be shrewd enough to discover'. Many are the crimes that have been committed in primary schools, when $(2 + 7) \times 3 - 8$ has been 'worked out' as '$2 + 7 = 9 \times 3 = 27 - 8 = 19$'. In the days when geometry involved formal proofs, there were pupils who started each line with the sign '=' (instead of the then-fashionable '∴'), hence using the verb '=' as a conjunction.† The same thing was (and still is) perpetrated in algebra

---

†When the use of '⇔' and its associated symbols first became fashionable in elementary work, the writer predicted that it would not be long before it too would have degenerated into yet another universal link, written down without any serious thinking taking place: he has seen many instances of this that have amply justified his apprehension. Perhaps it is the case that children *need* a sign to represent some vague connexion that they are not sufficiently articulate to express but, if that really is so, let us invent a special sign for that purpose and not keep grasping symbols that have already been invested with precise meanings and then allowing them to be used with anything less than perfect precision.

with solutions like

$$x^2 = 2x - 1 = x^2 - 2x + 1 = 0 = (x - 1)^2 = 0 = x - 1 = 0 = x = 1,$$

(which may be written on one line or on several). The reader should be able to sort this one out more easily than some examples of the art.

**Any use of '=' to mean anything other than 'is equal to' must be unreservedly deplored.**

It is the pupils who have been allowed to be sloppy in using the balancing sign 'is equal to' who have difficulty in solving equations; who, because they do not clearly perceive that '$3x - 5 = 7$' is making the absolutely precise statement that '$3x - 5$ *is equal to* 7', cannot readily proceed to '$3x - 5 + 5 = 7 + 5$' and so on.

Any teacher who has ever encouraged (or even condoned) the writing of foolishness like

'A bicycle = £35'   or   'Tom = 13 years'

instead of

'A bicycle costs £35'   and   'Tom is 13 years old'

must share some of the blame for all the difficulties children experience. Note that even writing

'Tom's age = 13 years',

although not as culpable as the previous monstrosity, is still NOT to be recommended: just write

'Tom's age is 13 years'.

A disservice is done to the cause of mathematical language every time the sign for 'is equal to' is devalued by such cavalier and unnecessary use.

Similarly, a statement such as

'the gradient when $x = \alpha$ is 3'

clearly becomes quite unacceptable if the pupil replaces 'is' by '=', and although (provided the writer can be trusted to have written gramatically)

'the gradient at $\alpha = 3$'

can have only one meaning, it is discourteous to subject the reader to that unnecessary jolt before he realizes that the sentence has ended. That is the sort of discomfort which, in other contexts, punctuation is designed to prevent. This analogy is not frivolous: the enclosure of a clause between commas or dashes or parentheses closes it up, so to speak, in relation to the other elements in the syntax; the presence of the sign '=' between $\alpha$ and 3 closes up the phrase '$\alpha = 3$' in a similar way. In the above statement, the unwanted bunching would have been avoided if 'is' had been used.

How then do you react to the student (or teacher) who writes

'when $x = \alpha$, the gradient $= 3$'   or   'at $\alpha$, the gradient $= 3$'?

These sentences must be acknowledged to be completely unambiguous but, even so, do you not feel that 'the gradient is 3' would have been incomparably better?

Teachers set poor examples to their pupils by unnecessarily replacing the simple word 'is' by a sign meaning 'is equal to', thereby encouraging their young charges to believe that the merit of a piece of mathematics is proportional to the number of occurrences per square metre of the sign '='. Mathematical symbols have evolved to promote concision of expression: in some hands, these servants seem to be in danger of becoming our masters.

It is not yet clear whether the trend in the early years of arithmetic towards writing things like

$$4 + 5 \rightarrow 9 \quad \text{or} \quad 4, 5 \overset{+}{\nearrow} 9$$

to mean that 'under the operation of addition, 4 and 5 combine to become 9' will have any advantages. Presumably, it is meant to prevent misuse of the sign of equality and to present addition as an active process rather than a passive state. The arrow is intended to suggest in a very elementary way a mapping (but without, of course, using that word). Since, however, what is being gently suggested is a function[†] $f$ of *two* variables, $f(a, b) = a + b$, with $a + b = c$ replaced by the idea that, under $f$, the number pair $(a, b)$ maps to $c$, that is, $(a, b) \overset{f}{\mapsto} c$, it may be felt to be a rather cumbersome approach. Time, however, will be needed to tell whether this interesting method does have any advantages in improving understanding and preparing the way for later work; and, in particular, whether, if the introduction of the sign '=' is delayed, its

---

†The concept of function and the various notations for expressing it are discussed in Chapter 2.

definition will be better appreciated and its correct use more stringently monitored.

Meanwhile, one of the saddest recent innovations in school mathematics has crept in as a by-product of the otherwise praiseworthy intention of introducing some computer programming into the syllabus. The enthusiasts for this development—surely they *cannot* have been practising teachers?—have written books actually suggesting that pupils should write

$$A = A + 3 \quad \text{or, alternatively,} \quad A := A + 3$$

to mean that the contents of location A are to be extracted, 3 is to be added and the result replaced in location A. Even the concessionary use of the colon in the second notation is utterly inadequate compensation for this gratuitous appropriation and misuse of the sign for 'is equal to'. As if teachers did not have enough of a struggle encouraging pupils to use language correctly, without having the ground cut from beneath their feet by misguided and insensitive authors!

Observe that it is yet another binary relation that is having to be symbolized, namely, 'is to be replaced by'. Happily, a few wiser writers have had the sense to introduce a more acceptable notation

$$A \leftarrow A + 3$$

to express this, but the harm has already been done.


## 1.4

Not all binary relations have their own private symbol and those that do not must be abbreviated (if at all) in other ways. Examples are

| | |
|---|---|
| 'is the negative of' | 'is the square of' |
| 'is the reciprocal of' | 'is the logarithm of' |
| 'is the inverse of' | 'is the complement of' (between sets or switches) |
| 'is the conjugate of' | 'is disjoint from' (between sets) |
| 'is coprime to' | 'is the negation of' (between propositions). |

Relations, of course, do not have to be binary, although these are the ones that are most likely to be given a place as a 'topic' in school work. Teachers, at least, should recognize that

> '. . . lies between . . . and . . .'
> '. . . is the difference of . . . and . . .'
> '. . . is the remainder when . . . is divided by . . .'

are ternary relations, whereas a proportion $(a : b = c : d)$ expresses a quaternary relation. A singulary† relation just identifies a property that an element of an appropriate set may or may not have, like 'is negative' (of real numbers), 'is prime' (of positive integers), 'is isosceles' (of triangles), 'is empty' (of sets), 'converges' (of series), 'is continuous' (of functions) and so on.

Notice that the use of the word 'relation' in the phrase 'recurrence relation' is in harmony with the general definition. This concept describes a particular type of relation among members of a sequence $(a_n)$, such as

$$F(a_{n+2}, a_{n+1}, a_n, n) = 0$$

or, when it has the special form called a *linear* recurrence relation,

$$a_{n+2} + p(n)a_{n+1} + q(n)a_n = r(n),$$

where F, $p$, $q$, $r$ are given functions. In this restricted context, however, the terminology is slightly different, as it has been adapted from that associated with differential equations. The above examples of recurrence relations are said to be *second order*, but, as you will see, they are *ternary* relations. A recurrence relation of order $m$ is, in fact, an $(m + 1)$-ary relation in the general sense. Recurrence relations, however, are not mentioned again in this book, except for some passing references in Chapter 18 on mathematical induction.

**1.5**

The general binary relation $\rho$ requires for its definition two sets S and T and the ability to say whether, given $s \in S$ and $t \in T$, the relation in question holds or not. If it does, one writes $s \rho t$. It is called a relation *between* S and T or, if T = S, a relation *in* S. The sets S and T themselves are not usually given names, but the subset of S defined by

$$\{s \in S : s \rho t \quad \text{for } some \quad t \in T\}$$

†The words 'binary', 'ternary', etc. are constructed from the following sequence of Latin distributive number adjectives:

singula, bina, terna, quaterna, quina, sena, septena, octona, novena, dena, undena, duodena, . . .

(these are the neuter forms), meaning 'one each' or 'one at a time', 'two each' or 'two at a time', and so on. This explains why purists favour 'denary' and 'octonary' when these members of the sequence are needed—rejecting the use of 'decimal' and 'octal' for that purpose—and why they prefer 'singulary' for the first member of the sequence, rather than the irregularly-formed adjective 'unary'.

may be called the *domain* of the relation and the subset of T defined by

$$\{t \in T : s \, \rho \, t \quad \text{for } some \quad s \in S\}$$

may be called its *range*. These names are in harmony with those used in connexion with functions (and explained in Chapter 2) but it is not absolutely necessary to introduce them in reference to a relation, whereas they are indispensable terminology for functions.

The properties of a binary relation $\rho$ in a set S that are most likely to be needed for elementary classification are the following, where $a$, $b$, $c$ denote elements of S.

| | | | | |
|---|---|---|---|---|
| Reflexive | For all $a$, | | | $a \, \rho \, a$ |
| Irreflexive | For no $a$, | | | $a \, \rho \, a$ |
| Symmetric | For all $a, b$, | $a \, \rho \, b$ | $\Rightarrow$ | $b \, \rho \, a$ |
| Antisymmetric | For all $a, b$, | $(a \, \rho \, b \, \& \, b \, \rho \, a)$ | $\Rightarrow$ | $a = b$ |
| Transitive | For all $a, b, c$, | $(a \, \rho \, b \, \& \, b \, \rho \, c)$ | $\Rightarrow$ | $a \, \rho \, c$. |

Note that the definition of 'antisymmetric' requires the *prior* definition of some relation of equality among the members of the set S and that, in the last two definitions, it is the logical conjunction (denoted here by &) of the two given propositions that implies (another relation!) the stated conclusion. A relation $\rho_0$ such that $a \, \rho_0 \, b$ for *no* $a, b$ is called the *empty relation* and a relation $\rho_1$ such that $a \, \rho_1 \, b$ for *all* $a, b$ is called the *universal relation*.

The two most important special categories of relation are, of course, the equivalence relations (which are reflexive, symmetric and transitive), and the partial order relations (which are reflexive, transitive and antisymmetric).

## 1.6

With any transitive relation, there are conventions to be observed in writing. Although the statement '$x \leq 2$ and $x \geq -1$' can be efficiently telescoped into '$-1 \leq x \leq 2$', the corresponding sort of simplification that pupils often try to concoct out of '$y \geq 2$ or $y \leq -1$', writing '$2 \leq y \leq -1$' is, of course, illicit, because the transitivity of the relation would then force the false conclusion that '$2 \leq -1$'. Note the significant fact that the conjunctions in the original compound statements are not the same: the first has 'and'; the second has 'or'.

There is, however, also a special convention regarding equivalence relations. This one is more subtle and is not always made clear to students. From a statement such as $a = b = c = d$, all sorts of deductions, such as $a = d$, $d = b$, $c = a$, can of course be made, because the relation is reflexive, symmetric and transitive. When used in *reasoning*, however, it is the transitive feature that is predominant. When

$$p = q = r$$

is written, the message being conveyed is that, from the data $p = q$ and $q = r$, the conclusion $p = r$ is being drawn. So, when a pupil has discovered that $t = u$ and that $t = v$ and writes $t = u = v$ to tell the reader that he wishes him to infer the fact that $u = v$, he is breaking this hidden convention. To achieve the result he wants, he must write

$$u = t = v.$$

Similarly, if he is deducing from $x = z$ and $y = z$ that $x = y$, and elects to collapse the argument into one phrase, he should write *not* $x = y = z$, but

$$x = z = y.$$

Unless students follow this accepted procedure, the logical structure of their arguments may be quite opaque. Consider, as a very simple example, the pupil who wishes to prove that each element in a group has a unique inverse: that is, that,

$$\text{if} \quad ax = xa = e \quad \text{and} \quad ay = ya = e, \quad \text{then} \quad x = y$$

($e$ being the identity element). Contrast the muddle exhibited in

$$(xa)y = ey = y = x(ay) = xe = x$$

with the smoothly-flowing argument in

$$x = xe = x(ay) = (xa)y = ey = y.$$

In less straightforward proofs, it can be extremely difficult for the reader to unscramble a student's reasoning and the writer himself may easily lose the thread of his own argument. The only alternative to adopting this standard convention is to insist that all the elementary statements are written separately and not linked together into one long chain but, in many proofs, that can be unnecessarily laborious.

**1.7**

There is another approved convention concerning the use of relation symbols. By permitting them to be used in subordinate clauses with a slightly modified meaning, considerable linguistic economy is achieved. In these contexts, the relation sign *still has a verbal function*—it is most important to realize this—but the appropriate reading may no longer be 'is . . .' and a relative pronoun may have to be supplied. A few representative examples of this type of usage follow.

| *Phrase* | *Meaning* |
|---|---|
| Given $\varepsilon > 0$, . . . | 'Given a number $\varepsilon$ that is greater than 0, . . .' |
| For all $a \in A$, . . . | 'For all elements $a$ that belong to A, . . .' or '. . . belonging to A, . . .' |
| There is an $n > N$ such that . . . | 'There is a number $n$, (which is) greater than N, such that . . .' |
| For every $d \mid n$, . . . | 'For every integer $d$ that divides $n$, . . .' or 'For every divisor, $d$, of $n$, . . .' |
| For some $A \supseteq B$, . . . | 'For some set A that contains B, . . .' or '. . . containing B, . . .' |
| There exists $w = \bar{z}$ such that . . . | 'There exists a complex number $w$, (which is) the conjugate of $z$, such that . . .' |
| When $x \notin \mathbf{Q}$ is given, . . . | 'When a number $x$, not belonging to $\mathbf{Q}$, is given, . . .' or '. . . an irrational number, $x$, . . .' |
| We can find a $\delta \geq 0$ such that . . . | 'We can find a number $\delta$, (which is) greater than or equal to 0, such that . . .' or '. . . find a non-negative $\delta$ such that . . .' |
| Let $z \neq 1$ be a complex number. | 'Let $z$ be different from 1 and a complex number.' or '. . . a complex number other than 1.' |
| Take $\mathbf{a} \perp \mathbf{b}$. | 'Take the vector $\mathbf{a}$ to be perpendicular to $\mathbf{b}$.' |
| Writing $A = \ln B$ requires $B > 0$. | 'Writing $A = \ln B$ requires $B$ to be positive.' |

The convenience of this extension of the symbolic language and the saving it affords will be obvious to you. It is, however, accepted that this luxury will only be vouchsafed to a single relation at a time: never abuse the concession by trying to write, for example, 'Given $A \neq B \subseteq C$, . . .' or

'For every $d \mid n < D, \ldots$' because of the misunderstandings that would inevitably arise.


## 1.8

The above treatment of relations is certain to be the one preferred for school use, because it puts the general relation symbol $\rho$ in the place where the binary relation symbols cited in the earlier list all go, namely, *between* the objects being related.

There is, however, a more abstract and clinical definition of a relation, which tends to be favoured by lecturers in colleges and universities. You see, from a more advanced standpoint, the trouble with the sort of presentation described so far is that it never quite defines what a relation actually is. It assumes that, as a result of previous experience, the reader will know the sort of entity being described and will recognize that he has to be able to say whether, given $s \in S$ and $t \in T$, the relation in question 'holds' or not. But precisely what this notion of 'holding' involves intuition is left to determine.

This lack of an absolutely watertight definition is remedied by presenting relations in a different way as subsets of the cartesian product $S \times T$ of two sets S and T. The cartesian product, you will recall, is the set of all ordered pairs of elements, one from S and the other from T. Thus

$$S \times T = \{(s, t) : s \in S, t \in T\}.$$

Now, the argument goes, there are some ordered pairs $(x, y)$ for which a given relation $\rho$ 'holds' (for which $x \, \rho \, y$), and others for which it does not. It is the aggregate of favourable pairs that is the real object of study. Conversely, any given set of ordered pairs determines a relation by the requirement that $x \, \rho \, y$ if and only if $(x, y)$ belongs to that set. [Note that complementary relations are associated with complementary subsets of $S \times T$, the empty relation with the empty set and the universal relation with the whole cartesian product $S \times T$.]

So a fresh start is made by *defining* a relation $\rho$ between two sets S and T to *be* a subset of $S \times T$; and $x \, \rho \, y$ can then be written to indicate that $(x, y)$ is an element of this subset $\rho$. But the trouble with this is that the sign $\rho$ is then being used in two *different* ways, and one is likely to be faced with bizarre statements like '$x \, S \times T \, y$' for '$(x, y) \in S \times T$' (the universal relation) or see the empty relation symbolized by '$x \, \varnothing \, y$' using the noun $\varnothing$ as a verb! Consequently, when this ordered pair approach is adopted, it is much more likely that all statements like '$x \, \rho \, y$' will be abandoned in favour of '$(x, y) \in \rho$'.

The verbal character of the symbol $\rho$ has then vanished: $\rho$ has become just a set. Although it is now much better defined, this antiseptic treatment has robbed $\rho$ of all the vigour of the verbal phrases in the list. And it was these traditional relations that provided the motivation for the study of general relations in the first place!

Moreover, one does not want any pupil to be replacing (even mentally) '$l \parallel m$' by '$(l, m) \in \parallel$' or 'A $\supseteq$ B' by '(A, B) $\in$ $\supseteq$'. The familiar symbols can under *no* circumstances be stretched to serve in that way and so a new letter has to be introduced for every such relation before it can be studied.

Indeed, for school presentation, the definition of a relation as a set is not at all attractive and the more lively exposition, with $\rho$ used as a verbal symbol, offers much better prospects for fruitful study. But teachers should understand where the topic is likely to lead, so that they can avoid saying anything that will make it unduly difficult for their ex-students to accept this very abstract definition of a relation if they meet it later.

**1.9**

A compromise treatment is, however, possible and this can be recommended for both school and college use; it will be rejected only by lecturers who are implacable fundamentalists. It starts with a relation between S and T firmly specified by a traditional verbal symbol $\rho$ (so that any reservations you or he may have about the nebulous idea of a relation 'holding' or 'not holding' do have to be suppressed). But, when the relation $\rho$ has been given, the associated subset of the cartesian product S $\times$ T is called the *graph of the relation* $\rho$ and is denoted *not* by $\rho$ but by, say, $\Gamma_\rho$. Thus

$$(x, y) \in \Gamma_\rho \quad \Leftrightarrow \quad x \, \rho \, y$$

or

$$\{(x, y) : x \, \rho \, y\} = \Gamma_\rho \subseteq S \times T.$$

This means, of course, that, in the case when S and T are both **R**, the set of real numbers, the 'graph of the relation $\rho$' may mean either this set $\Gamma_\rho$ of ordered pairs $(x, y)$ or the cartesian graph in which these same pairs $(x, y)$ are plotted as a set of points in the euclidean plane. But this is just the sort of fuzziness one actively encourages in coordinate geometry and should not cause misunderstanding.

Conversely, a given subset $\Gamma$ of S $\times$ T is called *not* a relation but the *graph of a relation* and a verbal symbol such as $\rho_\Gamma$ is introduced to denote the actual relation; that is,

$$x \, \rho_\Gamma \, y \quad \Leftrightarrow \quad (x, y) \in \Gamma.$$

This hybrid treatment may not satisfy all the demands of the purist, but it goes a long way towards harmonizing the two approaches. It certainly seems preferable to replacing '$x \, \rho \, y$' by '$(x, y) \in \rho$', using the *same* symbol $\rho$ and the *same* name 'relation' for both ideas (and thereby suggesting, for example, that '$(a, A) \in \in$' is considered a legitimate alternative to '$a \in A$'). That cannot be regarded as good teaching (or lecturing), however dedicated a set-enthusiast you are.

### 1.10

The discussion will now turn from an examination of relations to operations. In contrast to a relation, which has this verbal structure, a binary operation behaves grammatically more or less like a conjunction. In order to appreciate the importance of this fundamental distinction, you should inspect the list of common operations in Table 1.1 and carefully consider the part each plays in a mathematical sentence.

The binary operations mentioned in this table are all symbolized by what may be called *infixes*: signs placed *between* the objects whose combination is to be effected, as $x \circ y$. [The symbol '$\circ$' is often used to denote a general, unspecified, binary operation, as well as the particular operation of composition.] Not all binary operations have an infix symbol. One outstanding exception is exponentiation, where $b^a$ is written instead of employing an infix (although it is interesting to note that computer languages, which require an infix symbol, sometimes denote this power by $b \uparrow a$: see Chapter 9 on the notation for powers).

### 1.11

Any function of 2 variables defines a binary operation $\circ$ by

$$x \circ y = f(x, y).$$

Here the operation is being symbolized by a *prefix* symbol $f$. This is even clearer in the so-called Polish notation, introduced into mathematical logic by Jan Łukasiewicz. In this scheme, $p \wedge q$ is written $Kpq$, $p \vee q$ as $Apq$, $p \rightarrow q$ as $Cpq$, and so on. So, in this Polish notation, $p \vee [(p \rightarrow r) \wedge q]$ would be written as $ApKCprq$. Unlike an infix notation, a prefix notation does *not* require brackets, which is the

**Table 1.1**

| SET — between whose elements the operation is defined | BINARY OPERATION | | | | NOTES |
|---|---|---|---|---|---|
| | Name of operation | Symbol for operation | Reading of symbol | RESULT | |
| numbers, vectors, matrices | addition | + | plus | sum | Note that, associated with the arithmetic operations, is a complete set of names to distinguish between the operation * and the result $x * y$, as well as a separate word for reading the symbol * itself. |
| | subtraction | − | minus | difference | |
| numbers, matrices | multiplication | × (.) [or omitted] | [times] | product | |
| numbers | division | ÷ | divided by | quotient | |
| vectors | scalar multiplication | . [never omitted] | dot | scalar product | Unlike the dot for other types of multiplication, the dot denoting scalar multiplication is an integral part of the notation and must *never* be omitted. |
| | vector multiplication | ∧ (×) | vec; (cross) | vector product | The symbol for vector multiplication is unfortunately not the subject of any agreement.+ |

| | operation | symbol | word | set terminology | commentary |
|---|---|---|---|---|---|
| sets | cartesian multiplication<br>intersection operation<br>union operation<br>addition<br>set difference operation | $\times$<br>$\cap$<br>$\cup$<br>$+$<br>$\setminus$ | cross<br>cap; and<br>cup; or<br>plus<br>but not | cartesian product<br>intersection<br>union<br>sum<br>set difference | There is a lengthy examination of the notation and terminology for set operations in Chapter 15. Observe that the set of words for these set operations is *not* complete, so that, for example, the word 'union' may mean the result (A ∪ B) or sometimes the operation (∪) itself. |
| propositions | conjunction operation<br>disjunction operation<br>addition<br>conditionality<br>biconditionality | $\wedge$ (&)<br>$\vee$<br>$+$<br>$\rightarrow$<br>$\leftrightarrow$ | and<br>or<br>plus<br>only if<br>if and only if | conjunction<br>disjunction<br>sum<br>conditional<br>biconditional | The binary operations between propositions are discussed in detail in Chapter 16 and are included in this list for comparison and cross-reference. In this subject, the excellent general word 'connective' is used to describe an infix symbol and sometimes, by extension, to refer to the operation itself. |
| mappings, functions, permutations, geometric transformations, symmetry operations | composition | $\circ$ (.)<br>[or omitted] | [circle; of] | composite | The problem of deciding whether '$\sigma$ followed by $\tau$' shall be denoted by $\sigma \circ \tau$ or by $\tau \circ \sigma$ is dealt with in Chapter 14 on non-commutative operations. |

†Some early writers on vectors denoted scalar multiplication by $\times$ and others used **x** in the vector product, so that at one time the cross was ambiguous. Nowadays, there is no danger of that: the cross would only be used for vector multiplication. But there seems to be a strong case for using the distinctive wedge sign ($\wedge$) for vector multiplication. This has the further slight advantage for schoolteachers of leaving the cross free for multiplication of scalars: one may, for example, wish occasionally to write $|\mathbf{a} \wedge \mathbf{b}| = \frac{1}{2} \times 1 \times \sin \frac{1}{4}\pi$.

particular reason for its importance. But it is undeniable that their omission makes it fiendishly difficult to take in at a glance the meaning of a string of symbols. The idea of a Polish-type notation does not, of course, have to be restricted to operations in logic. It is perfectly possible to represent other operations by prefixes and to write, say, S$xy$ for $x + y$, D$xy$ for $x - y$, P$xy$ for $x \cdot y$, etc., so that DP$x$D$yz$Pz$S$xy$ would mean $x(y - z) - z(x + y)$. The bracket-free feature of Polish notation makes it convenient for certain computer applications. Indeed, it will be found that reverse Polish notation has been used in certain pocket calculators and so may be familiar to some students. Reverse Polish notation might write $x + y$ as $xy$S, for example, so that, using the same letters now as *postfixes*, the above algebraic expression would become, in reverse Polish notation, $xyz$DP$zxy$SPD. It is perhaps not entirely conservatism that leads one to prefer the more familiar infix notation in general, where the brackets, which then *have* to be used, help the eye to grasp the structures of expressions.

For a similar reason, the standard notation for the value taken by a function will probably always remain $f(x_1, \ldots, x_n)$, *with* the conventional parentheses, even though they are, strictly speaking, redundant. Not only are the brackets an aid to clarity but, in complicated expressions involving several functions, they remind the reader at each repetition of the number of variables associated with each function (or, if you prefer, the dimension of the domain of each function). For example, $fxgyxz$ means

$f(x, g(y, x, z))$   if $f$ is a function of 2 variables and $g$ of 3

$f(x, g(y, x), z)$                    3                    2

$f(x, g(y), x, z)$                    4                    1.

For school presentation, the parentheses in $f(x)$ must certainly be retained, if only because they emphasize the different roles the letters $x$ and $f$ play in the expression $f(x)$.

Ternary and more complicated operations are almost always represented by a prefix (functional) notation. So, often, are singulary operations; indeed, a function of a single variable can always be thought of as defining such an operation. But consideration of the notations used to represent the results of performing various singulary operations

$-x$   for the negative of $x$ ($x$ a number, vector or matrix)

$y^{-1}$   for the reciprocal or inverse of $y$

$A'$, $\bar{A}$ or $\complement A$   for the complement of a set $A$

$\neg p$, $\mathsf{p}$, $\sim p$, $\tilde{p}$, $\bar{p}$, $p'$ or N$p$   for the negation of a proposition $p$

$\sqrt{x}$   for the non-negative square root of $x$ ($x$ non-negative)

$$y^3 \quad \text{for the cube of } y$$
$$n! \quad \text{for factorial } n \text{ (a postfix notation)}$$
$$|t| \quad \text{for the modulus of } t$$
$$\mathscr{R}z \text{ or } \mathscr{X}(z) \quad \text{for the } x\text{-axal part of } z$$
$$\mathscr{I}z \text{ or } \mathscr{Y}(z) \quad \text{for the } y\text{-axal part of } z$$

will remind you of the variety of notation that is possible.

**1.12**

The features of interest for *operations* [conjunctions] are of necessity different from the features of interest for *relations* [verbs]. The properties of a binary operation $\circ$ that are most likely to be needed for elementary classification are listed below.

| | |
|---|---|
| Commutative | $a \circ b = b \circ a$ |
| Associative | $a \circ (b \circ c) = (a \circ b) \circ c$ |
| Idempotent | $a \circ a = a$ |
| Distributive (over a second operation $*$) | $a \circ (b * c) = (a \circ b) * (a \circ c).$ |

Strictly, the last definition is of left distributivity, of course.

It cannot be too strongly emphasized that the questions that may legitimately be asked about a relation or about an operation become meaningless when asked about an entity of the other sort. It makes sense to ask 'Is this binary relation symmetric?' and 'Is this binary operation commutative?': the questions become nonsense if the two adjectives are interchanged. There is no question of this type that can correctly be asked about both a relation and an operation.

So far, so good. But there is a snag, for which the inconsistency of mathematicians is responsible. If $f$ is a function of 2 or more variables, it is customary to say that $f$ is a symmetric function if, for any permutation $\sigma$ of the variables,

$$f(\sigma(x_1), \sigma(x_2), \ldots, \sigma(x_n)) = f(x_1, x_2, \ldots, x_n).$$

For example, if

$$f(\alpha, \beta, \gamma) = \lambda(\alpha^2 + \beta^2 + \gamma^2) + \mu(\beta\gamma + \gamma\alpha + \alpha\beta),$$

then $f$ is a symmetric function of $\alpha$, $\beta$, $\gamma$ for all constants $\lambda$, $\mu$. Pupils will meet this terminology when studying the symmetric functions of the roots of a polynomial equation. Although this use of the word 'symmetric' is mainly employed when there are 3 or more variables, it can be (and is) used when there are only 2. So, if $f(x, y) = f(y, x)$ for all $(x, y)$ belonging to the domain of $f$, you will see that $f$ may be called *either* a commutative function *or* a symmetric function. [For example, in a vector space V over the real field **R**, it may be possible to define an inner product $\phi$: this is an extension of the idea of a scalar product for geometric vectors. A mapping $V \times V \rightarrow R : \phi$ is called an inner product if it satisfies 5 axioms, one of which requires that $\phi(x, y) = \phi(y, x)$ for all $x, y \in V$. It is usual to describe this axiom by saying that the inner product has to be 'symmetric'. A better description would be 'commutative', but it is because of dilemmas like this that confusion spreads.] Although the binary *operation* o defined by $x \text{ o } y = f(x, y)$ for such a function $f$ is always called commutative (and *not* symmetric), the close association of these two words in this context makes it inevitable that students will have difficulty in appreciating the distinction emphasized in the preceding paragraph.

### 1.13

There is another general property of a binary operation whose importance is sometimes underestimated. Suppose an operation that is in some sense analogous to multiplication is defined on a set A in which an obvious candidate for being a zero element has already been identified. If this 'multiplication' has the property that

$$a \cdot 0 = 0 = 0 \cdot a$$

for all $a \in A$, an important question always to decide then is whether there are any *zero divisors*, that is, any solutions of

$$x \cdot y = 0$$

for which $x \neq 0$ and $y \neq 0$.

If $p$ and $q$ are real numbers, of course, pupils will recognize from experience that

$$p \cdot q = 0 \quad \Rightarrow \quad (p = 0 \text{ or } q = 0)$$

and there are no zero divisors, but it is not self-evident that the same property will, for instance, hold when $p$ and $q$ are allowed to become complex.

In finite arithmetics with a composite modulus there *are* zero divisors; for example,

$$3 \cdot 4 \equiv 0 \quad (6).$$

Also, in the system of geometric vectors† with vector multiplication as the operation,

$$(a = 0 \quad \text{or} \quad b = 0) \qquad \Rightarrow \qquad a \wedge b = 0,$$

but the converse implication is false because, of course,

$$a \wedge b = 0 \quad \text{whenever} \quad a \parallel b.$$

[Be careful, incidentally, in work with vectors, always to distinguish the *zero vector* **0** from the *zero scalar* 0.]

Similarly, in matrix algebra,

$$(A = O \quad \text{or} \quad B = O) \qquad \Rightarrow \qquad AB = O,$$

but a zero matrix O *does* have zero divisors; for example,

$$\begin{bmatrix} 2 & -4 \\ -5 & 10 \end{bmatrix} \begin{bmatrix} 20 & 8 \\ 10 & 4 \end{bmatrix} = \begin{bmatrix} 0 & 0 \\ 0 & 0 \end{bmatrix}. \ddagger$$

It is for this reason that, from a matrix equation such as

$$(A - \lambda_1 I)(A - \lambda_2 I) \ldots (A - \lambda_n I) = O,$$

it can *not* be inferred that

$$A = \lambda_r I \quad \text{for some} \quad r.$$

[Note that, just as the symbol for a unit matrix is a capital letter I, so also the symbol for a zero matrix is a capital *letter* O, and consequently any zero matrix should always be *read* as 'oh'.]

This possible existence of zero divisors explains why, when complex numbers are introduced, it is essential for students to prove that there are none in **C**. Until this point is satisfactorily established, it is not possible to deduce from

$$(z - \lambda_1)(z - \lambda_2) \ldots (z - \lambda_n) = 0$$

that

$$z = \lambda_1, \quad \text{or} \quad \lambda_2, \ldots, \quad \text{or} \quad \lambda_n,$$

---

†In abstract vector spaces, multiplication of vectors is not generally defined.
‡It is possible, as in this example, to choose A and B so that both AB = O and BA = O; it is also possible to arrange that just one of these relations shall hold.

as one needs to when solving a polynomial equation in complex algebra. It is, of course, a standard theorem that no field has zero divisors, but at the stage when complex numbers are first introduced, the general axioms for a field will not have been considered.

When complex numbers are defined as ordered pairs of real numbers, it is quickly discovered that $(0, 0)$ is the natural zero element in the new system and that

$$(x, y) \cdot (0, 0) = (0, 0) = (0, 0) \cdot (x, y)$$

for all $(x, y) \in \mathbf{C}$. So the relation to investigate is

$$(a, b) \cdot (c, d) = (0, 0).$$

The simplest way to deal with this condition is to observe that

$$(u, v) = (0, 0) \qquad \Leftrightarrow \qquad u^2 + v^2 = 0,$$

since $u \in \mathbf{R}$, $v \in \mathbf{R}$. Thus

$$
\begin{aligned}
(a, b) \cdot (c, d) = (0, 0) &\Leftrightarrow (ac - bd, bc + ad) = (0, 0) \\
&\Leftrightarrow (ac - bd)^2 + (bc + ad)^2 = 0 \\
&\Leftrightarrow a^2c^2 + b^2d^2 + b^2c^2 + a^2d^2 = 0 \\
&\Leftrightarrow (a^2 + b^2)(c^2 + d^2) = 0 \\
&\Leftrightarrow a^2 + b^2 = 0 \quad \text{or} \quad c^2 + d^2 = 0 \\
&\qquad \text{since} \quad \mathbf{R} \text{ has no zero divisors} \\
&\Leftrightarrow (a, b) = (0, 0) \quad \text{or} \quad (c, d) = (0, 0).
\end{aligned}
$$

[Later, of course, this conclusion will be seen to follow more shortly from (1) $z = 0 \Leftrightarrow |z| = 0$ and (2) $|z_1 z_2| = |z_1| \cdot |z_2|$, which is what the above proof is saying in more elementary language.]


### 1.14

Having already signposted several of the pitfalls for the teacher, it is advisable to point out finally that there are at least two other uses of the word 'operation' in school mathematics, which may, superficially at least, seem to be different from the ones considered so far.

The first is the use of the word 'operation' to describe a geometric transformation such as a rotation, reflexion, translation, glide reflexion, enlargement, shear or inversion. These transformations are mappings applied either to a general point in the plane (or in space), or to a pattern or rigid body. [In the latter case, the particular transformations being studied will probably be those called the symmetry *operations* of the figure: those mappings that yield an image superposable on the original figure.] All such geometric transformations, however, are really nothing other than singulary operations. Their novelty is that they are applied to geometric objects, but they are not conceptually different from other singulary operations.

If the transformations always map collinear points to collinear points they are called *linear*: all the examples in the above list are linear except inversion in a circle or sphere. [But the operation that maps $(x, y, z)$ to $(-x, -y, -z)$, sometimes called 'central inversion', *is*, of course, linear.] When, in addition, the origin is a fixed point of the transformation, the mapping (or function) is often called a (linear) *operator* and the same name may, by association, be applied to a matrix representing such a linear transformation relative to a given basis. The word 'operator' can, however, be regarded as an unnecessary extra item of jargon; there are plenty of synonyms for it.

Note particularly that, when the composition of geometric transformations is investigated, pupils are faced with operations of *two* kinds: the operation (binary) of composition is being applied to certain mappings, which are themselves operations (singulary).

The other use occurs when describing an operation performed on an expression (such as doubling it, squaring it, or deriving or integrating it wo† a specified variable), or on a function (such as obtaining the inverse function or the derived function). Here also, however, there is not really any new idea. The 'operations' described are just singulary operations in the earlier sense (just like those that double a number, or square it or take its reciprocal). The operation this time, however, is being performed on a function (or an expression) to give another function (or expression), rather than on a number to yield another number. In this context also, the word 'operator' appears, to describe $d/dx$ or $\int \ldots dx$.

## 1.15

Obviously, any discussion like the one in this chapter comes late in a student's education. Relations and operations are not first taught as 'topics': particular examples are studied as they arise and their relevant properties noted. Normally a relation and an operation will not be taught together; pupils should not have to study more than one of these concepts at a time. Varied experiences with both ideas will be accumulated gradually and only at a fairly late stage will attempts be made to draw the threads together by looking at a general relation or at a general operation. Any deliberate initiation of the sort of discussion

†With respect to.

above would come still later (unless, of course, it arose spontaneously from children's questions), since each idea needs to be implanted firmly and separately before a useful comparison can take place without the danger of confusion.

But all these background considerations must be present in the teacher's mind, however elementary the lesson, so that good foundations can be laid. It is because the precepts discussed here have not always been observed that many treatments of boolean operations and relations are, in the writer's opinion, seriously defective. This thesis is argued in Chapters 16 and 17.

# Chapter 2

# Functions

---

## 2.1

The idea of a function has undergone a lengthy process of development and refinement over the centuries and yet the best modern definitions are not only crisper, they are actually conceptually simpler than many of their once popular predecessors, which have been discarded during this evolutionary process. Changes in the definition of mathematical concepts eventually—although often painstakingly slowly—filter down into school textbooks. In this case, because of the clarification that has resulted from the stripping away of inessential and unhelpful accretions, modern ideas on functionality can be presented to pupils surprisingly early in their mathematical careers.

Vestiges of earlier treatments often linger on, however, even of ones that were fashionable in the last century. You may, for instance, hear a teacher or lecturer call '$x^2 - 5x + 2$' a function, or say 'Let $w$ denote the function $ax + by + cz$', or talk about a 'many-valued' function. The tenacity of these old habits is well illustrated by our lingering attachment to the phrase 'complementary *function*' in the theory of linear differential equations.

Textbooks do not always live up to the promise offered by the precise general statements their authors make in the early chapters. One touchstone by which you can often judge what a writer *really* understands by a function is the way he handles the inverse trigonometric functions: sometimes what he says there belies his own impressive definitions. This will be examined later. (Another indicator—if he goes that far—is his treatment of the logarithm of a complex number.)

## 2.2

As with so many other concepts, it was in the work of Euler that the notion of a function began to crystallize and it is to him that the notation $f(x)$ is due. To Euler, however, the word 'function' was more or less synonymous with a formula involving $x$, such as a power of $x$, a polynomial, or a rational or trigonometric expression, by means of which $y$ was defined. He does, however, also seem to have had the idea—quite a useful one for that era—that any drawn graph could be considered to define a more general sort of function.

It was, however, soon realized that such naive definitions led to difficulties and could not be accepted uncritically. This sparked off a lengthy period of debate among mathematicians about these fundamental questions, which was effectively brought to an end by Lejeune Dirichlet in 1837. He finally cut loose from the restraints imposed by the belief that formulae had to be lurking somewhere in the background to the definition and boldly said that $y$ is a function of $x$ if, when $x$ is given, a unique value of $y$ is determined.† This devastatingly simple idea blew away all the fog in a single gust and cleared the way for the developments in analysis in the latter half of the nineteenth century.

If you look at textbooks that were in favour in the first half of the present century, you will find definitions that say things like the following.

1. $y$ is said to be a function of $x$ if there is a relation between $x$ and $y$ such that, when the value of $x$ is known, the values of $y$ are also known.

2. A quantity $y$ is called a function of another quantity $x$ if values of $y$ are determined by values of $x$.

3. A variable $y$ is called a function of another variable $x$ if, with each value of $x$, is associated one or several values of $y$, according to some definite rule.

4. A function is a formula involving $x$ and having a definite value when any specific value is given to $x$.

5. When one quantity $y$ depends on another quantity $x$ in such a way that $y$ assumes a definite value when a definite value is given to $x$, $y$ is called a function of $x$: the variable $x$ to which a value is given is called

---

†Dirichlet's 1837 definition (below) is actually of a continuous, rather than a general, function and at that time, as you will observe, the concept of continuity was rather rudimentary and kinematic. But in other respects the modern criteria for a function are introduced.

'Now if, to each $x$, there corresponds a unique, finite $y$ in such a way that, as $x$ steadily traverses the interval from $a$ to $b$, $y = f(x)$ varies in a similarly progressive way, then $y$ is called a continuous function of $x$ for this interval. It is moreover not at all necessary for $y$ to depend on $x$ in accordance with the same law over this whole interval; indeed, the dependence does not have to be envisaged as expressible by mathematical operations. If presented geometrically, with $x$ and $y$ as abscissa and ordinate, a continuous function appears as a connected curve for which only one point corresponds to each abscissa between $a$ and $b$.' G Lejeune Dirichlet, Repertorium der Physik, Bd 1, 1837, 152–3.

the independent variable and the function itself is called the dependent variable.

6. A variable $y$ is called a function of the variable $x$ when, to any value of $x$, there corresponds a definite value of $y$.

The first three sorts of definition are pre-Dirichlet in outlook: they allow several values of $y$ to correspond to one value of $x$. What they are trying to characterize would (in modern parlance) *not* be called a function, but a (many-many) relation.

The last three types of statement all mention 'a definite value' and so make a concessionary nod in Dirichlet's direction (although any reference to a formula, as in (4), is harking back to the eighteenth century). But it was quite common for books that gave definitions like (4)–(6) to go on immediately to offer as 'functions' examples such as

$$f(x) = x \pm \sqrt{x},$$

which do *not* satisfy Dirichlet's—or that author's—definition, and to say that relations such as

$$y^2 + 3xy - x^2 = 5 \quad \text{or} \quad \tan y = x$$

implicitly define $y$ as '*a* function' of $x$. There was usually much discussion about so-called 'one-valued' and 'many-valued' (or 'single-valued' and 'multiple-valued') functions, with the better authors at least having the grace to point out that it was only for 'one-valued functions' that a derivative could be defined.

What such writers had failed to appreciate was Dirichlet's other brilliant innovation in his definition. Apart from his disengagement from the earlier reliance on formulae for the specification of functions, Dirichlet had added the simple requirement that, for any given $x$, the associated $y$ should be *unique*. The significance of this improvement took a surprisingly long time to percolate down into the elementary books, yet its merits are obvious. For, suppose that it *is* possible for $f(x)$ to be, say, $x \pm \sqrt{x}$, $(x > 0)$, and that the derivative of $f(x)$ is to be found. This requires first the evaluation of

$$\frac{f(x + h) - f(x)}{h}$$

and, when the above statement about $f(x)$ is accepted *at its face value*, this quotient is

$$\frac{h \pm \sqrt{(x + h)} \pm \sqrt{x}}{h}$$

with a double ambiguity of sign. For two of the four choices of sign, the quotient does not tend to a limit as $h \to 0$. It is manifestly *not* possible to let the symbol $f(x)$ be 2-valued in this way. It is a matter of paramount importance *which* value is being taken. Either $f(x)$ can be $x + \sqrt{x}$ exclusively or $f(x)$ can be $x - \sqrt{x}$ exclusively; what $f(x)$ cannot, under any circumstances, be is $x \pm \sqrt{x}$ indifferently.

The leeway that 'many-valued' functions apparently gave for obtaining $dy/dx$ implicitly when $y$ was determined as '*a* function' of $x$ by a relation such as $ax^2 + 2hxy + by^2 = 1$ was illusory. An implicit function theorem—please consult a textbook of analysis for this—is required to establish the possibility of *defining* a suitable ('one-valued') $f(x)$ from such a relation *before* any derivative $f'(x)$ can be satisfactorily invested with meaning.

Because, in defining derivatives and integrals and in applications such as series expansions, it is *essential* for the symbol $f(x)$ to be unambiguous, it is obviously far more satisfactory, as Dirichlet recognized, to *build into the definition* of a function the requirement of 'single-valuedness', and this is the normal practice nowadays:

**a 'many-valued' function is a self-contradictory concept.**

[Regrettably, in complex analysis, the other, antiquated point of view has not entirely disappeared and you can still find recent books that claim, for instance, that, if $w^2 = z$ or $\exp w = z$, then $w$ is 'a function' of $z$. (The relation $\exp w = z$ is discussed in some detail in Chapter 8). In complex analysis, no less than in real, $f(z)$ must be 'one-valued' by definition, otherwise, for example, the existence of a derivative $f'(z)$ cannot be sensibly investigated. You may perhaps know that the topological idea of a Riemann surface avoids this issue in a particularly elegant way: but that is a much more advanced idea.]

## 2.3

How then did these earlier authors cope with the inverse circular and hyperbolic functions? It was common practice to show that the 'function', $\sin^{-1}$, defined by making

$$'y = \sin^{-1} x' \qquad \text{equivalent to} \qquad 'x = \sin y'$$

had the property that

$$\frac{d}{dx}(\sin^{-1} x) = \pm \frac{1}{\sqrt{(1 - x^2)}},$$

and then to say that, because of the wish to use this 'function' in finding $\int dx/\sqrt{(1 - x^2)}$, the restriction $-\frac{1}{2}\pi \leq y \leq \frac{1}{2}\pi$ would be imposed on $y$ in order to be able to assert that

$$\frac{d}{dx}(\sin^{-1} x) = \frac{1}{\sqrt{(1 - x^2)}} \qquad (|x| < 1).$$

[Remember that the sign '$\sqrt{\phantom{x}}$' means 'the non-negative square root of'; this is reviewed in Chapter 3.] The writers that talked about 'many-valued' functions would say that '$x = \sin y$' defined $y$ as a 'many-valued function' of $x$, which had a so-called 'principal value' satisfying $-\frac{1}{2}\pi \leq \sin^{-1} x \leq \frac{1}{2}\pi$. It was this 'principal value' that was to be understood in the statements

$$\int \frac{dx}{\sqrt{(1 - x^2)}} = \sin^{-1} x + A \qquad (|x| < 1)$$

and

$$\sin^{-1} x = x + \frac{1}{2}\frac{x^3}{3} + \frac{1 \cdot 3}{2 \cdot 4}\frac{x^5}{5} + \frac{1 \cdot 3 \cdot 5}{2 \cdot 4 \cdot 6}\frac{x^7}{7} + \ldots \qquad (|x| \leq 1).$$

Thus the symbol '$\sin^{-1}$', as used by these authors, was always ambiguous. Echoes of this treatment may still be heard in classrooms even today. The confusion it used to cause is well illustrated by a sequence of examination questions from around the 1950s, quoted in Section 7.2. These are, after all, sufficiently recent to represent the sort of way that teachers now practising were being taught when they were at school.

A few writers at least saw the merit of having a more sensitive notation and significantly improved their presentation by using $\text{Sin}^{-1}$ (with an upper case S) for their so-called 'many-valued function' and $\sin^{-1}$ (lower case s) for their 'principal value'. They would regard the whole curve in Fig. 2.1 as the cartesian graph of $y = \text{Sin}^{-1} x$ and the part represented by the solid line as the graph of $y = \sin^{-1} x$. This was a considerable notational improvement, but their conceptual treatment is not acceptable today because, as explained above, a function *must* be what these older authors used to call 'one-valued', otherwise it does not qualify as a function. Thus, according to the modern canon, only their $\sin^{-1}$ (*not* $\text{Sin}^{-1}$) is a function. [That is not to say that there is not an interesting *relation* lurking in the background: this is considered in Chapter 7.]

Note that the problem is not merely one of an ambiguous sign. This

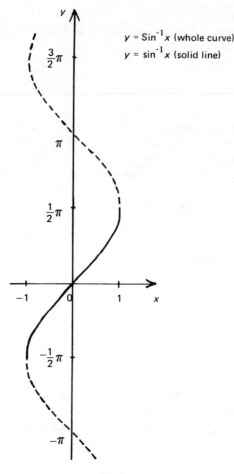

Fig 2.1

can be seen with $\text{Tan}^{-1}$. Here the 'principal value', $\tan^{-1}$, was defined by $-\frac{1}{2}\pi < \tan^{-1} x < \frac{1}{2}\pi$ and it is, of course, this $\tan^{-1}$—it cannot be $\text{Tan}^{-1}$—that occurs in

$$\int \frac{dx}{1 + x^2} = \tan^{-1} x + A$$

and in

$$\tan^{-1} x = x - \frac{x^3}{3} + \frac{x^5}{5} - \frac{x^7}{7} + \ldots \qquad (|x| \leq 1).$$

Observe also that there was no ambiguity with $\mathrm{sh}^{-1}$, since '$x = \mathrm{sh}\, y$' necessarily determines a unique $y$ for any given $x$, but with '$x = \mathrm{ch}\, y$', a similar distinction had to be created between $\mathrm{Ch}^{-1}x$ and $\mathrm{ch}^{-1}x$, as illustrated in Fig. 2.2.

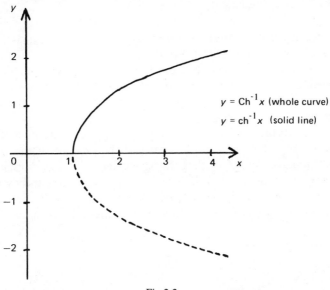

$y = \mathrm{Ch}^{-1}x$ (whole curve)

$y = \mathrm{ch}^{-1}x$ (solid line)

Fig 2.2

## 2.4

At this stage in the development of the concept, the only functions being considered were functions of a real (or complex) variable. For many of these, particularly those given by formulae, there was some intrinsic restriction on the values that $x\ (\in \mathbf{R})$ or $z\ (\in \mathbf{C})$ could take unless additional definitions were made. For example,

$$f(x) = x^{-1} \quad : \ x \neq 0 \qquad\qquad f(x) = \ln x \quad : \ x > 0$$

$$f(x) = \sqrt{(x^2 - 1)} : \ x \geqslant 1 \text{ or } x \leqslant -1$$
$$f(z) = \frac{1}{z^2 + 1} \ \therefore \ z \neq i, -i$$
$$f(x) = \cot x \quad : \ x \neq n\pi, n \in \mathbf{Z}$$

and so on. But the tendency was to insist that $x$ (or $z$) should be allowed to take all real (or complex) values apart from the ones that were 'obviously' exluded. This, as will be seen later, corresponds to the viewpoint of coordinate geometry rather than of analysis. The fact that a less dogmatic attitude might be helpful becomes clear when the composition of functions is considered. If, for example, the $x$ in sin $x$ is allowed to take all real values, then ln sin $x$ is not well-defined. To invest ln sin $x$ with meaning, $x$ must be confined to values for which sin $x > 0$, that is, to the union of the intervals†

$$(2n\pi, (2n + 1)\pi) \quad \text{for} \quad n \in \mathbf{Z}.$$

In other words, it is often necessary to restrict the set of values taken by $x$ (or $z$) to some subset of $\mathbf{R}$ (or $\mathbf{C}$) and, even if not necessary, the freedom to do this may be most useful. So the advantage of incorporating this flexibility into the definition of *any* function, even when there is no 'natural' requirement for such a limitation, comes to be recognized.

Thus a function is now being defined on a given subset of $\mathbf{R}$ (or $\mathbf{C}$) and that set of values of $x$ (or $z$) is called the *domain* of the function. Gradually the idea emerges that the specification of the domain is the first and most fundamental task in defining any function, and that, if the domain is changed, then the function is changed, even if the 'rule' for determining the value of the function is the same in both cases.

### 2.5

There are still several more developments to record before the concept reaches its modern form. In the older treatments, it was $f(x)$ that was called a function. Immense clarification was achieved as soon as mathematicians began to distinguish between $f$ and $f(x)$, and to call $f$, rather than $f(x)$, the function. This led to the idea of $f$ being treated as a *mapping*, which operates on $x$ to yield $y$. Thus $y$, or $f(x)$, denotes the *result* of this procedure, that is, the *image* of $x$ under the mapping $f$,

---

†The convention of using

$(a, b)$ for the open interval $\{x \in \mathbf{R} : a < x < b\}$

and

$[a, b]$ for the closed interval $\{x \in \mathbf{R} : a \leqslant x \leqslant b\}$

is followed in this book. The alternative notation of $]a, b[$ for the open interval is quite satisfactory and distinctive; it has the great advantage of removing all possibility of confusion between the open interval $(a, b)$ and the ordered pair $(a, b)$ or the greatest common divisor $(a, b)$. But, for some reason, the present writer happens to find this ingenious notation jarring and distracting whenever he sees it in print, which is why he has chosen not to adopt it here. The use, however, of $<a, b>$ for the open and $\leqslant a, b \geqslant$ for the closed interval has absolutely nothing to commend it and should be shunned by teachers: remember that '$<$' and '$\leqslant$' are verbs and you should never write anything that suggests otherwise.

whereas the function $f$ becomes identified with the *process* and *not* with the end-product. It was presumably the rise of abstract algebra that created the climate in which the importance of such a distinction could be appreciated. For, in subjects like group theory and matrix algebra, operations such as permutations and geometric transformations had become objects of study in their own right. The raw material of such investigations was often a collection of operations that transformed one mathematical entity into another, not those entities themselves. This disentanglement of $f(x)$ and $f$ seems, in retrospect, to have provided the key to a clear understanding of functionality, the goal towards which mathematicians had been groping for so long.

### 2.6

The task now, therefore, is to delineate the precise requirements associated with the notion of a mapping. But it is just as easy to do that after a further generalization has been introduced. This liberates the concept of 'function' from an exclusive dependence for its domain on a set of real or complex numbers (even though in practice such sets are always likely to provide the most important domains). Indeed, appreciation that functional dependence is not, fundamentally, an analytic idea at all, but involves merely the existence of a mapping from one *general* set to another yielded further valuable insight.

So now one starts with 2 sets, D and C, called respectively the *domain* and *codomain* (or target) of the function, and a mapping $f$ that associates with *each* element $x \in D$ a *unique* image $f(x) \in C$. Notice the 2 essential properties demanded by this definition:†

(1) $f(x)$ is defined for *every* $x \in D$,
(2) $f(x)$ is *unique*, so that

$$x_1 = x_2 \quad \Rightarrow \quad f(x_1) = f(x_2).$$

These statements are saying respectively that

(1) each element of D is linked to *at least* one element of C
(2) each element of D is linked to *at most* one element of C,

and so together they require that

each element of D is linked to *exactly* one element of C.

†You should be warned that, if you search the literature, you will be able to find a few authors that have chosen to omit one or other of these properties from *their* definition of a mapping. But, as you will discover, there are powerful reasons for including them both and defining a mapping in the above way.

The words 'function' and 'mapping' can (certainly with children) be treated as synonymous, but there *is* a delicate difference: a function is really 'a domain *together with* a mapping'. Having specified the domain D, the mapping is the process by which the elements of D are transformed into elements of C. But a function is not just a disembodied mapping; one should never forget that the specification of the domain is the first imperative in the definition of a function. [A pedant might even insist on different letters being used for the mapping and for the function!]

The admission of a completely general set as a domain for a function means that the concept can now be introduced much earlier in the curriculum by using small finite sets for D and C, rather than starting with **R** or subsets of **R**. The conditions inherent in a mapping can be made clear by simple diagrams like Fig. 2.3.

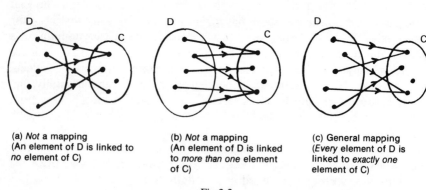

(a) *Not* a mapping (An element of D is linked to *no* element of C)

(b) *Not* a mapping (An element of D is linked to *more than one* element of C)

(c) General mapping (*Every* element of D is linked to *exactly one* element of C)

Fig 2.3

## 2.7

Until recently, writers tended to use the ordinary arrow '→' for 'maps to' but, thanks partly to encouragement by the Open University, the more distinctive barred arrow '↦' has come into wider use. The concept of mapping is so important that the advantage of having a special notation exclusively reserved for 'maps to' is overwhelming. But, apart from that, the simple arrow '→' is in established use in analysis for 'tends to' and, although admittedly the occasions when actual confusion might arise will be rare, it is more stylish to have a symbolism that can instantly distinguish between

$$x \mapsto 3 \quad (x \text{ maps to } 3)$$

representing a constant function, ($f(x) = 3$ for all $x \in D$), and

$$x \to 3 \quad (x \text{ tends to } 3)$$

used in the ordinary sense.

If $f$ denotes the mapping under which $x$ maps to $y$, the standard notation is

$$f : x \mapsto y. \tag{1}$$

There is, however, a modification that is sometimes useful when finding images under a chain of successive mappings. This puts the mapping letter above the mapping arrow, writing

$$x \overset{f}{\mapsto} y \overset{g}{\mapsto} z$$

to mean that, under $f$, $x$ maps to $y$ and that, under $g$, $y$ maps to $z$. This symbolism is often convenient, for example, in transformation geometry and in the study of permutations.

A mapping statement like

$$f : x \mapsto x^3 + 4x$$

can, if you like, be read as '$f$ maps $x$ to $x^3 + 4x$'. But a *better* reading is '$f$ [pause] $x$ maps to $x^3 + 4x$', since this puts the whole phrase 'maps to' in the place where the arrow occurs.

The definition of the mapping rule, however, is only one of the *two* pieces of information about a function that need to be presented symbolically. The other is a mention of the domain D and codomain C of the function. This is usually expressed by writing

$$f : D \to C \tag{2}$$

and here an *ordinary* arrow *is* used, thereby helping to emphasize the distinction between (1) and (2). Indeed, the use of the differently shaped arrows to represent these two ideas has been a revolutionary advance in the notation.

Usually, when defining a function, both types of statement have to be made. (Exceptions arise if the context permits some sort of blanket announcement that all functions about to be introduced have a certain domain D and codomain C.)

The writer would like to offer for consideration the new suggestion that the second type of statement above should *always* be written the other way round as

$$D \to C : f.$$

This can be read as 'D into C [pause] $f$'.

Then, when both assertions have to be made, they can be elegantly telescoped into

$$D \to C : f : x \mapsto f(x);$$

for example,

$$\mathbf{R} \to \mathbf{R} : f : x \mapsto x^3 + 4x,$$

which can be read as '$\mathbf{R}$ into $\mathbf{R}$ [pause] $f$ [pause] $x$ maps to $x^3 + 4x$'. If either component does not need mentioning, it can be suppressed *without altering* the order of the rest of the specification.

It must be stressed that this is *not* established practice, but there do not seem to be any snags. Quite the contrary: one big advantage that can be claimed for this impovement is that the *first* thing mentioned in the full prescription is the *domain*, giving the correct emphasis to the whole theory. The notation in this form will be used experimentally throughout this book.

## 2.8

At one time, the codomain was sometimes called the range of the function, but nowadays a most important distinction is maintained between these two terms. The *codomain* (or target) denotes *any* set to which *all* the images under $f$ of elements of $D$ belong. It may (and usually does) contain elements that are not the image of any element of D, since it is usually quite unnecessary to go to the trouble of finding the smallest set for the codomain.† For example, when defined as

$$\mathbf{R} \to \mathbf{R} : F : x \mapsto x^2 + x + 3,$$

the codomain is just given as $\mathbf{R}$; this is quite adequate. It could have been taken instead to be the set of positive real numbers, or the set

---

†Although the *concept* of the codomain is important—essential, even—the present writer is not fully persuaded that *mentioning* a codomain *in the notation* is such a good idea, although he has followed the fashion and done so in this book. This insistence on including the codomain in the specification seems to stem from a belief that a function is *merely* a many-one relation; this attitude is going to be examined presently.

$\{x \in \mathbf{R} : x > 2\}$. All that is essential is that it contains the set $\{x \in \mathbf{R} : x \geq 11/4\}$, which happens to be the set of all images of the elements of the domain.

It is this complete set of images under $f$ of elements of $D$ that is called the *range* of the function $f$. [You will notice that this definition does not conflict with that of the range of a *relation*, mentioned in Section 1.5.] The range of $f$ is the set

$$\{f(x) : x \in D\}$$

and may be written $f(D)$. In a schematic mapping diagram, it may be represented as a subset of C (Fig. 2.4). Note that it is preferable *not* to talk about the image of a function. One speaks of the *image of an element* of the domain, or the *image of any subset* of the domain; but one speaks of the *range of the function* (or of the mapping).

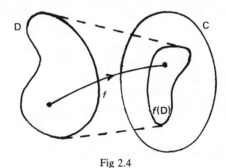

Fig 2.4

## 2.9

It is an unfortunate linguistic accident that the word chosen, 'range', begins with the letter 'R'. English-speaking mathematicians seem to have a penchant for giving their concepts names beginning with that letter! The notation for the Real and Rational numbers is resolved by using **R** for the Real and **Q** for the Rational numbers (suggesting 'Quotients'). But that still leaves the Rings that one usually denotes by R; the general Relations that one sometimes wants to symbolize by R (although, because of this pressure on the letter R, $\rho$ comes to the rescue as a useful alternative); the (quadratic) Residues for which R is normally used; and the Ranges of functions that one would have liked to denote by R had R not been so heavily oversubscribed already. Then there is the use of R for Reflexive (applied to a relation), R for a Radius (of convergence, or otherwise), R (as in $R_n$) for a Remainder, R for a Rotation and R for a Reflexion!

The problem, as you will know, is sometimes eased by the use of open capitals

$$\mathbb{Z}, \mathbb{Q}, \mathbb{R}, \mathbb{C}$$

for the systems of integers, rational, real and complex numbers respectively—this also solves the dilemma of C standing for the complex numbers and for a codomain—with the addition usually of $\mathbb{N}$ for the natural numbers $\{0, 1, 2, \ldots\}$ and perhaps $\mathbb{P}$ for the positive integers $\{1, 2, 3. \ldots\}$ as well. This distinctive notation is well established in advanced work, but some teachers find it rather fussy for school use. In this book, bold face letters

$$\mathbf{P, N, Z, Q, R, C}$$

are being used for these sets of numbers.

There are, incidentally, strong technical reasons why most mathematicians prefer to *include* zero among the natural numbers; there is the additional practical point that the term is superfluous in elementary work if 'natural number' is synonymous with 'positive integer'. The technical reasons, for those of you that are interested, concern the introduction of the natural numbers as cardinals of sets. When a natural number $n$ is the cardinal of a certain set S, then the successor number $n'$ is defined to be the cardinal of the set $S \cup \{S\}$. It is 'natural' to start this inductive process with the empty set, and to define zero as its cardinal Thus

the natural number 0 is the cardinal of $\varnothing$

| | |
|---|---|
| 1 | $\{\varnothing\}$ |
| 2 | $\{\varnothing, \{\varnothing\}\}$ |
| 3 | $\{\varnothing, \{\varnothing\}, \{\varnothing, \{\varnothing\}\}\}$ |

and so on.

## 2.10

On the occasions when the set C chosen for the codomain is actually the range of the function, the mapping is then said to be a mapping *onto* C, and it is called a *surjective* mapping or *surjection*. In this case, there is a further bit of symbolism that can be employed, although its use is *not* obligatory. To emphasize, when necessary, that C is actually the range of $f$, one can write

$$D \twoheadrightarrow C : f,$$

read as 'D *onto* C [pause] $f$'. This statement implies that $C = f(D)$.

Having mentioned the word 'surjection', which you may regard as an item of rather highbrow and trendy jargon, it must be explained that it is already coming into school use through certain textbooks. This is a development you may or may not applaud, but it is rather important for you as a teacher to make yourself familiar with the terms 'surjective', 'injective' and 'bijective' applied to mappings, ready for the day when you feel confident enough to experiment with them in your lessons—or find yourself being asked to do so.

A mapping $D \rightarrow C : f$ is called an *injective* mapping (or *injection*) if it is one-one; that is, if

$$f(x_1) = f(x_2) \quad \Rightarrow \quad x_1 = x_2.$$

A mapping that is both surjective and injective is said to be *bijective* or to be a *bijection*. In older books you will find a bijection described as a 'one-one correspondence' between D and C. All these ideas can again be illustrated (Fig. 2.5) by taking manageable finite sets for D and C.

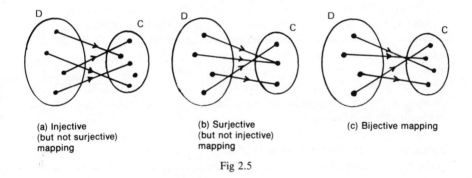

(a) Injective
(but not surjective)
mapping

(b) Surjective
(but not injective)
mapping

(c) Bijective mapping

Fig 2.5

For an *inverse* function $C \rightarrow D : f^{-1}$ to be definable, $f$ *must be bijective*; because, if $f$ is not surjective, then at least one element of C has no partner in D and, if $f$ is not injective, then at least one element in C has more than one partner in D. This necessary condition for the existence of an inverse function is obviously also sufficient. These observations and Fig. 2.5(a), (b) should be matched with Fig. 2.3(a), (b) and the remarks about mappings just before those diagrams. These correlations are very important.

## 2.11

It may be, however, that you never met these technical terms in your mathematics courses; perhaps your lecturers talked about funny things called epimorphisms and the like. This will be a good place to try and disentangle for you the two systems of nomenclature and to clear up any confusion you may have experienced if you met both.

The idea of a mapping from one set to another is a very general one and takes no account of the nature of the elements themselves. If the sets actually have some structure (if they are, say, groups or fields or vector spaces), then the *interesting* mappings are the ones that preserve that structure.

Let $\circ$ be the infix symbol for a binary operation, defined for every pair of elements in D, and with respect to which D is closed (that is, $a \circ b \in D$ whenever $a \in D$ and $b \in D$). If C is closed wo a corresponding binary operation $\square$ and $D \rightarrow C : f$ has the property that, for all $a, b \in D$,

$$f(a \circ b) = f(a) \square f(b),$$

then $f$ is said to preserve this operation. Similar definitions can be formulated for singulary operations, ternary operations and so on.

When both sets D and C are structures of the same type (for example, both rings or both boolean algebras), then a mapping that preserves *all* the operations inherent in the structure of D is called a *morphism*. (This was once, and sometimes still is, called a homo-morphism.) In other words, a morphism must do two things: (1) it must be a mapping from D into C with all the restrictions that are necessary for it to qualify as such, and (2) it must preserve all the relevant algebraic structure.

A morphism that is also a surjection (a mapping onto C) is called an *epimorphism*; a morphism that is also an injection (a one-one mapping) is called a *monomorphism*; a morphism that is both an epimorphism and a monomorphism (a morphism that is a bijection) is called an *isomorphism*.

This is clearer if presented in tabular form.

| mapping | | morphism |
|---|---|---|
| surjection | onto | epimorphism |
| injection | one-one | monomorphism |
| bijection | onto and one-one | isomorphism |

[You will observe that the right column is constructed from Greek roots, whereas the left column is of basically Latin origin, but has been subjected to French hybridization. Incidentally, as far as the writer is aware, nobody has yet suggested calling a mapping a 'jection'.]

Finally, there is a further item of optional extra jargon. In the case when C is the same as D (or is a subset of D), a morphism of D into itself is called an *endomorphism* and, if C = D, an endomorphism that is also an isomorphism is called an *automorphism*.

### 2.12

You will probably find it a relief to turn now from these abstract definitions to matters of more immediate classroom interest. The recognition that a function is a mapping has given the teacher a lively new technique for presenting functions to quite young children, which was impossible with the earlier and vaguer definitions.

A function is imagined as a machine, with an input and an output, whose operation mimics the mapping process. When an appropriate input (corresponding to $x$) is fed in, the machine produces an output (corresponding to the image $f(x)$); diagramatically,

$$x \qquad \boxed{\quad f \quad} \qquad f(x)$$

.

Although the machine is being treated as a 'black box', the hidden mechanism must be deterministic: feeding in the same input must always produce the same output. This is to ensure that the mapping condition

$$x_1 = x_2 \quad \Rightarrow \quad f(x_1) = f(x_2)$$

is satisfied.

When teaching, one probably starts with machines that do simple arithmetic operations on numbers, such as

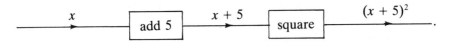

.

The contrast between this composite machine and

$$x \qquad \boxed{\text{square}} \qquad x^2 \qquad \boxed{\text{add 5}} \qquad x^2 + 5$$

already raises important issues for pupils to discuss. You can progress later to machines with 2 (or more) inputs to simulate binary (and more complicated) operations. Machines can also be imagined that perform operations in geometry, logic, etc, as well as arithmetic mappings.

The other idea essential to the concept of a function, its domain, can be developed from the notion (mentioned above) that the input has to be 'appropriate'. Choosing the domain is equivalent to assigning to the elements belonging to that set a coded key† that will cause the machine to operate. Whenever an input element $x$ has the correct code associated with it, the machine produces the output $f(x)$: if an element without that code is fed in, the machine will fail to operate and there will be no output.‡ (This corresponds to the other mapping condition.)

You will by now appreciate the dazzling clarity with which the idea of a function can be illuminated in this way once it has been stripped to its bare essentials. It is pleasant to speculate that Euler, who was an imaginative innovator, would probably be delighted if he could return and see the way in which his first tentative ideas on functionality have been steadily refined and polished until they can be presented in this playful and mechanistic way. It reinforces mathematicians' belief that the subject is basically simple, once it has been properly understood. One wonders how many other mathematical ideas that now seem impossibly abstruse to all but the specialist are really only complicated because they are surrounded by extraneous clutter, or because our present viewpoint is unhelpful, and will be capable of similar clarification as soon as we have wit enough to penetrate to the core of the concept.

### 2.13

The inverse trigonometric functions§ are no problem to the present-day teacher. He merely ensures that the domain of the *original* function is restricted in such a way that the resulting function is *bijective*. It has already been explained that this is a necessary and sufficient condition on *any* function for it to have an inverse. So the definition of these particular inverse functions becomes merely a special case of a quite general process, rather than an apparently ad hoc procedure to facilitate integration.

---

†Compare the idea of the characteristic function of a set, discussed in Section 16.6.

‡Please make sure you are distinguishing *this* response from that in which the machine operates and produces a zero output.

§As a matter of convenience, rather than of strict logic, it is useful to use *'trigonometric functions'* as a portmanteau term to include all circular and hyperbolic functions.

Thus it is the *bijective* functions

$$\left.\begin{array}{rcl} [-\tfrac{1}{2}\pi, \tfrac{1}{2}\pi] & \twoheadrightarrow & [-1, 1] \quad : \sin \\ (-\tfrac{1}{2}\pi, \tfrac{1}{2}\pi) & \twoheadrightarrow & \mathbf{R} \quad\quad : \tan \\ \{t \in \mathbf{R} : t \geq 0\} & \twoheadrightarrow & \{t \in \mathbf{R} : t \geq 1\} : \mathrm{ch} \end{array}\right\} \text{ that have the inverses}$$

$$\left\{\begin{array}{rcl} [-1, 1] & \twoheadrightarrow & [-\tfrac{1}{2}\pi, \tfrac{1}{2}\pi] \quad : \sin^{-1} \\ \mathbf{R} & \twoheadrightarrow & (-\tfrac{1}{2}\pi, \tfrac{1}{2}\pi) \quad : \tan^{-1} \\ \{t \in \mathbf{R} : t \geq 1\} & \twoheadrightarrow & \{t \in \mathbf{R} : t \geq 0\} : \mathrm{ch}^{-1}. \end{array}\right.$$

A complete set of inverse circular and hyperbolic functions can be defined, but those above, together with

$$\mathbf{R} \twoheadrightarrow \mathbf{R} : \mathrm{sh}^{-1}$$
$$(-1, 1) \twoheadrightarrow \mathbf{R} : \mathrm{th}^{-1},$$

suffice for all normal purposes (integration, series expansions, etc). The cartesian graphs of these inverse functions are drawn in Fig. 6.1, to illustrate some incidental remarks about integration in Chapter 6. Note also that it is these same inverse functions, defined as above, whose values ($\sin^{-1} x$, $\tan^{-1} x$, etc) occur as the sums of various standard power series (see Section 2.3 and Chapter 13).

**2.14**

So much, then, for the presentation of a function as a domain together with a mapping. There is, however, another 'modern' definition of function, which is also very popular. This one does not mention the word 'mapping', but defines a function as a special type of relation. This apparent economy of concepts makes this approach, on the face of it, very attractive. But it is the same false sort of economy as that which sets out to confuse $x \, \rho \, y$ with $(x, y) \in \rho$ [see Section 1.8], and, indeed, is much favoured by authors who do just that.

If a relation $\rho$ between S and T has the property that

$$\text{if} \quad s \, \rho \, t_1 \quad \text{and} \quad s \, \rho \, t_2, \quad \text{then} \quad t_1 = t_2,$$

(in other words, if no $s \in S$ is related to more than one $t \in T$), then $\rho$ is called a *many-one* relation.[†] Several writers then go on to say that 'a many-one relation is called a function'. But this is a most inadequate definition. The main snag is that there is no mention of a domain for the function.

---

†As used here, 'many-one' will be assumed *not* to exclude 'one-one', just as 'rectangle' does not exclude 'square'.

In Section 1.5, the domain of the relation $\rho$ is defined as the set

$$D = \{s \in S : s \,\rho\, t \quad \text{for some} \quad t \in T\}.$$

Since this is a well-defined set for any $\rho$, it may seem as though this procedure will also provide a satisfactory explanation of the domain of a function. But, for any relation, many-one or otherwise, this idea of its domain arises very much as a postscript, only introduced *after* $\rho$ has been defined, whereas for a function the emphasis has to be reversed, making the *given* domain the primary concern.

For a general relation $\rho$, $D \subset S$ is allowed: indeed, D is usually a proper subset of S. Moreover, for a function, it is only D that is important: the original set S has no significance in this case. So, since S is the given set for defining a relation, whereas D is the given set for defining a function, the only hope of reconciling the two concepts arises when $D = S$. It would, however, be highly undesirable if the definition of every relation $\rho$ were to be fettered by the requirement that $D = S$. Yet, for a function, where the starting point *must* be D, that is exactly what *is* required. This is why it is misleading simply to say that any many-one relation between two given sets is called a function. (To make this clear, observe that the graph in Fig. 2.3(a) represents a many-one relation between D and C but *not* a function from D into C.) Note also that, although the sets S and T form an ordered pair, and are hence distinguishable, their roles with respect to $\rho$ are virtually identical,† in marked contrast to the roles of D and C in the definition of a function.

Given sets D and C, a relation $\rho$ between D and C must therefore have *two* properties if it is to define a function from D into C:

    (1) D must be the domain of $\rho$,
    (2) $\rho$ must be a many-one relation.

(Note how these correspond to the two conditions for a mapping, discussed earlier.) So, if you favour this relational approach to functions, you should check your reference books carefully to make sure that they include both requirements.

---

†It is because of this fact that one does not do *too* much violence to the language by talking about a relation 'between' S and T, despite the fact that they do not enter completely symmetrically and so 'between S and T' has to be distinguished from 'between T and S'. On the other hand, a mapping 'between D and C' would be an intolerably loose description: here the phrase '*from* D *into* C' is mandatory.

**2.15**

But, even though a relation having both properties *defines* a function, is it reasonable to say that such a relation *is* a function? Logically, one can: the concept is well-defined. But it is rather like the definition of a relation as merely a set, which came in for criticism in Section 1.8. All the special stimulus provided by the mapping concept has been dissipated. Furthermore, the notation used is different; the symbol $f$ is neither a verb (like $\rho$) nor a set (like $\Gamma_\rho$). For a function, one writes '$f : x \mapsto y$' or '$y = f(x)$', *not* '$x f y$', in the same way one would write '$x \rho y$' for a relation. (Admittedly '$x \overset{f}{\mapsto} y$' comes pretty close to this, but the verb is contained in '$\mapsto$', not in '$f$'.)

The arrow emphasizes the dynamic nature of the mapping process, which is the essential novelty of the function concept. It is an idea necessarily absent from that of a relation. The statement '$x \rho y$' is a vivid description of a property enjoyed by certain $x$ and $y$, but that property is intrinsically passive: it is not an active transformation converting $x$ into $y$, as a function is.

Also, of course, one does *not* write '$(x, y) \in f$', although one can find certain 'a-function-is-a-relation-is-a-set' fanatics who take apparent pleasure in doing so.

**2.16**

The difference between a relation and a function may be further pinpointed by looking at their possible graphic representations, and by the significant observation that

*coordinate geometry involves a study of relations: analysis involves a study of functions.*

Cartesian graphs are plotted in both these branches of mathematics, but the different purposes for which they are drawn need to be sharply contrasted. The distinctions may seem slight, but conceptually they are very important. They can be exposed by presenting the graphs in different ways in the two subjects.

The object of interest in 2-dimensional coordinate geometry is a *binary relation* in **R** and the set of ordered pairs $(x, y)$, $x \in$ **R**, $y \in$ **R**, between which that relation holds. For example, if the relation is

$$y^2(a - x) = x^2(a + x) \quad (a > 0),$$

then the set of all pairs that satisfy this relation is represented by the points of a certain curve, which is called the *graph of the relation*. When plotted, points satisfying this relation† may conveniently be represented by *crosses* (Fig. 2.6), until a sufficient number has been obtained to enable the curve to be sketched. The curve in this case depicts the complete set of points satisfying the relation. [The curve in Fig. 2.6, in case you are interested, is known as a strophoid.]

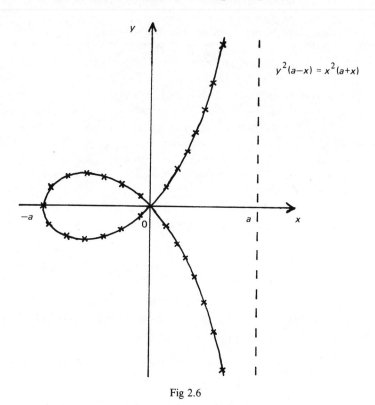

$$y^2(a-x) = x^2(a+x)$$

Fig 2.6

There is no mention in the definition of any bounds on the values of $x$ and $y$. It is easily discovered that, for this particular relation, $-a \leqslant x < a$ whereas $y$ may take any value. These sets are the domain and range of the relation. As has just been emphasized, these features of a *relation* emerge *after* the definition has been made: they do not have to be specified in advance. It is always understood that, in default of any contrary indication, *the graph of a relation occurring in coordinate geometry is to include every point* $(x, y)$ *with* $x \in \mathbf{R}$, $y \in \mathbf{R}$ *that satisfies the relation.* This ties in exactly with the other use of the term 'graph of

---

†A loose, but convenient, shorthand for 'points representing number pairs that satisfy this relation'.

a relation' (see Section 1.9) to mean the subset of $\mathbf{R} \times \mathbf{R}$ consisting of all number pairs $(x, y)$ for which the relation holds.

Similarly, the graph of the relation

$$\frac{x^2}{a^2} + \frac{y^2}{b^2} < 1$$

is the set of points in the interior of a certain ellipse and the graph of the relation

$$x^2 + y^2 + 1 = 0$$

is the empty set.

The object of interest in analysis is a *function*: a given domain, together with a particular mapping from that domain into a suitable codomain. Here *the domain is specified in the definition* and is of paramount importance. The functions

$$[-1, 1] \to \mathbf{R} : g : x \mapsto 3x^2 - 1$$
$$\mathbf{R} \to \mathbf{R} : G : x \mapsto 3x^2 - 1$$

are *different* functions. The fact that the mapping is performed according to the same rule is irrelevant; the functions are *automatically* different when the domains are different. (Changing the codomain, of course, does *not* change the function, although both codomains must be admissible sets.) Similarly,

$$[-1, 1] \to \mathbf{R} : h : x \mapsto \sqrt{(1 - x^2)}$$

is a well-defined function, but

$$\mathbf{R} \to \mathbf{R} : H : x \mapsto \sqrt{(1 - x^2)}$$

does *not* define a function, because no image $H(x)$ is specified when $|x| > 1$. (Recall the elementary examples in Fig. 2.3.)

When functions are plotted using a *mapping graph*, arrows are drawn from points belonging to a set representing the domain to their image points in a set representing the codomain. For example, Fig. 2.7 shows a representation of the function $g$ above by such a mapping graph. (The axis lines do not have to be parallel—or even coplanar—although they can be if you wish.) Such graphs are excellent for emphasizing the mapping definition but they cannot convey in a single glance the overall behaviour of a function in the same way that a cartesian graph can.

But even for a *cartesian graph*, it is *not* the set of images
$\{f(x) : x \in D\}$ that needs emphasizing: it is still the mapping $x \mapsto f(x)$;
the *process* of getting *from x to f(x)*. To highlight this, a function $f$
ought to be represented by *arrows* drawn perpendicular to the $x$-axis,
each arrow starting at a point $x$ in the set representing D (a subset of
the $x$-axis) and ending at the point with ordinate $f(x)$. This is illustrated
in Fig. 2.8, which displays the same function $g$. To join up the tips of the
arrows, however tempting, would be to *obscure* the essential feature of
this representation, which imagines such an arrow drawn from *every*
point of the domain. A function graph is a collection of arrows, *not* a set
of points. The joined *arrow tips*, therefore, have *no* significance for the
graph of a *function* in analysis, unlike the joined *crosses* marking a curve
associated with a *relation* in coordinate geometry.

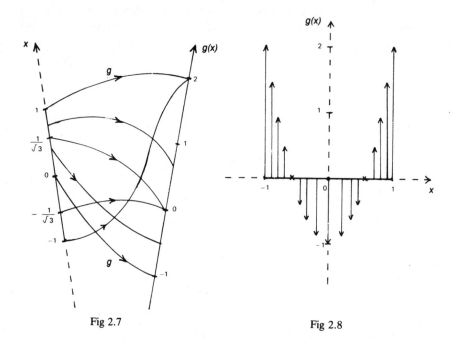

Fig 2.7                                          Fig 2.8

It may be objected that the arrows in Fig. 2.8 do not join any
logically chosen points (in contrast to the arrows in the mapping graph,
Fig. 2.7). They are nevertheless the natural vehicle for encoding the
information when the function is depicted by a plot in the euclidean
plane. They use the metrical property of the length of the arrow (and its
sense) to convey the value of the image, $g(x)$, as well as stressing the
mapping process, $x \mapsto g(x)$, by the arrows themselves.

**2.17**

This distinction between a relation and a function would be easier to bring home to pupils were it not for our practice of using the letter $y$ for the dependent variable in analysis. It would be much better if some other letter, such as $w$, were *regularly* used for this purpose, even when there is only one independent variable, $x$. In geometry, the coordinates $x$ and $y$ that occur in a binary relation (or the $x, y, z$ in a ternary) do so on an exactly equivalent footing. This is quite different from the way in which the dependent variable ($w$, say) in analysis is distinguished from the independent variable(s) $(x, y, z, \ldots)$. It is the use of the *same* letter, $y$, for these very different purposes—for one of the coordinates in geometry and for the dependent variable in analysis—that makes the confusion so difficult to unscramble and is probably the reason why so many readers and listeners are prepared uncritically to accept claims that a function is *merely* a many-one relation. Sadly, however, the whole weight of analytic tradition is against any such clarification of the two subjects by using some letter *other than* $y$ for the image of $x$ under a function whose domain is a subset of **R**. Having stated the case, therefore, no further effort to swim against this particular current will be made. (For instance, in Chapter 10, a strong temptation to use $w$ for $f(x)$ and to replace $dy/dx$ by $dw/dx$ throughout was successfully resisted!)

**2.18**

Whenever a cartesian graph is drawn, either the geometric or the analytic aspect is predominant. If the rectangular hyperbola is being studied in geometry, it is the *relation $xy = 1$*, represented by Fig. 2.9, that is important. If the integral of $1/x$ is being investigated in analysis,

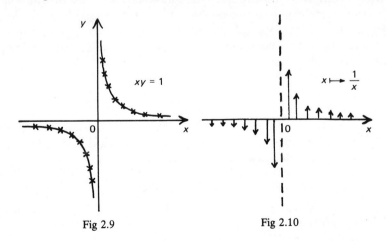

Fig 2.9                                    Fig 2.10

it is the *function* with domain $\mathbf{R} \setminus \{0\}$,† the non-zero real numbers, defined by the mapping $x \mapsto 1/x$, that is under consideration and the cartesian graph *should* then be presented as in Fig. 2.10.

These different types of plot emphasize the different purposes for which cartesian graphs are drawn in the two subjects. Confusion of these objectives does not help the pupil to develop clear ideas on relations and functions. But it must, of course, be admitted that it would take real dedication to persevere with this distinction. Regrettably, most teachers are likely—rather quickly—to capitulate and content themselves with substituting for the cartesian graph of the *function* $D \to \mathbf{R} : f : x \mapsto f(x)$, the cartesian graph of the associated many-one *relation* $y = f(x)$, by plotting the number pairs $(x, f(x))$ for $x \in D$ with crosses rather than arrow tips and, horror of horrors, joining them up! Let us not delude ourselves, however. By doing this, we are deliberately extinguishing a delicate but crucial distinction and, once quenched, that spark may be difficult to rekindle.

### 2.19

One of the consequences of the modern presentation, with its discrimination between $f$ and $f(x)$, is that mathematicians have left themselves without a good name to describe $f(x)$—as opposed to $f$— and they do not seem to feel any urgent need to coin a suitable term. When $D \subseteq \mathbf{R}$ and $a$ is one specific element of D, one can say 'the value $f(a)$'. But in teaching one often wants a general name for $f(x)$, thinking of it as an aggregate of values for all $x \in D$. To talk about 'the function $f(x)$' would be to demolish at a stroke the edifice one has been painstakingly erecting.

Often, certainly, one can usefully and correctly speak of 'the image $f(x)$'. But it feels slightly uncomfortable to talk about 'integrating an image' or 'squaring an image', although perhaps the passage of time will make such statements acceptable. It seems to the writer that, in *present* circumstances, the *least* unsatisfactory solution when teaching is to call $f(x)$ an *expression*. The trouble with this suggestion, as you will have spotted at once, is that it rather strongly conjures up again the discarded idea of $f(x)$ being some specific algebraic or trigonometric formula and may, therefore, be felt to destroy all Dirichlet's good work. At first sight, for example, it seems an inappropriate description of $\psi(x)$ when $\mathbf{R} \to \{0, 1\} : \psi$ is Dirichlet's function, defined by

$$\psi(x) = 1 \qquad \text{if } x \text{ is rational}$$

†The notation 'A \ B' ('A but not B') for the set of all elements of A that do not belong to B is discussed in Section 15.5.

$$\psi(x) = 0 \qquad \text{if } x \text{ is irrational.}$$

Nevertheless, if the image $\psi(x)$ is thought of as *'expressing the result'* of performing the mapping $\psi$ on $x$, the terminology may perhaps not seem too unreasonable. *It is certainly far less objectionable than calling $f(x)$ a function, which should be avoided at all costs.*

Similarly, one should be on one's guard against the trendy author, who thinks he is at the forefront of modern fashion when he defines a sequence as 'a function whose domain is **N**' (or **P**). What nonsense! It is surely

$$s(0), s(1), s(2), \ldots, s(n), \ldots$$

that is the sequence, *not* the function $s$.

In the special case of polynomials, this language problem does not arise. If

$$p(x) = c_0 + c_1 x + \ldots + c_n x^n,$$

then the expression $p(x)$ is a *polynomial*, the function $p$ is a *polynomial function*. If $c_n \neq 0$ and $n = 1, 2, 3, 4, \ldots$, $p(x)$ is a linear polynomial, a quadratic (polynomial), a cubic (polynomial), a quartic (polynomial), $\ldots$; $p$ is respectively a linear function, a quadratic function, a cubic function, a quartic function, $\ldots$. Notice, in passing, that, if $p(a) = 0$, it is customary to say that '$a$ is a *root* of (the polynomial *equation*) $p(x) = 0$', but that '$a$ is a *zero* of (the *polynomial*) $p(x)$'. [Also, of course, that quaint phrase '$p(x)$ vanishes at $a$' may be used.]


**2.20**

There is, however, another, closely related, general problem of terminology. If one has an explicit functional relation, $y = f(x)$, this can be satisfactorily read, in the usual way, as '$y$ is equal to $f$ (of) $x$'. On occasions, however, one may wish to make a general statement to the effect that this type of relation exists between $x$ and $y$ without mentioning an actual symbol for the mapping $x \mapsto y$. What one usually says then is that '$y$ is a function of $x$'. If you think about this claim, you will see that it is really complete nonsense; $y$ is patently *not* a function: $y$ is an element of the codomain of a mapping. A more precise statement would be '$y$ is functionally related to $x$', but that is less concise. *If* you feel that it is too much of a mouthful to say this—and many people obviously do—perhaps the best solution in this case is grudgingly to allow the traditional statement '$y$ is a function of $x$',† but to encourage plenty of discussion to bring out the fact that this phrase is just a conventional shorthand for 'there is a mapping with a suitable domain under which $x$ maps to $y$'.

To reject the assertion that '$f(x)$ is a function' and yet to accept (albeit reluctantly) statements like '$y$ is a function of $x$' or '$w$ is a function of 3 real variables' is not, one hopes, mere logic-chopping. There does seem to be an important, if subtle, linguistic nuance to be detected here. In rather the same spirit, it is sometimes convenient (as in Chapter 6) to speak of something as being 'a function of an angle' or to say, as a scientist is very likely to do, that 'the speed of the particle is a function of the time' or 'the volume of the gas is a function of the pressure and the temperature'.

It will, however, probably take us a long time to educate all our science colleagues to distinguish between $f$ and $f(x)$ and *not* to call $f(x)$ a function. But then, it has taken us a long time to reach our present understanding of this concept. Moreover, we cannot really claim that we have fully assimilated the modern definition ourselves, as long as we continue to call certain expressions 'complementary *functions*'.

†But do *not* say '$y$ is equal to a function of $x$'.

# Chapter 3

# Square roots

---

## 3.1

Pupils are usually intrigued by square roots. Part of this fascination is doubtless due to their discovery that positive numbers have two square roots and negative numbers have none. This statement refers, of course, to the system of numbers the children are familiar with at the time, although there is the additional complication that they may also pick up from the layman the impression that 'the square root of minus one' (together with 'centrifugal force', which he also does not understand properly), is some sort of magic key that unlocks the secrets of the universe for mathematical initiates. Also, of course, having probably been already told that $\pi$ is 'equal' to $\frac{22}{7}$ (or, in these days, even told that $\pi$ is 'equal' to 3), a number like $\sqrt{2}$ may be the pupils' first real experience of a decimal, $1 \cdot 41421 \ldots$, that lacks a pattern.

Some of the mystery surrounding square roots then attaches to the sign '$\sqrt{\ }$', which they rather enjoy using, regarding it as a powerful talisman. So impressed are they that sometimes even older and otherwise sensible students insist on taking square roots at every possible opportunity, rushing to write down $\sin x$ as soon as $\sin^2 x$ has been obtained, or $dy/dx$ as soon as they see $(dy/dx)^2$, however great the complexity thereby introduced. Experienced teachers will confirm for you that it is a useful indication of a pupil's mathematical maturity if he instinctively refrains from rewriting his equation in a form where the sign '$\sqrt{\ }$' has to be used, unless it is absolutely essential.

## 3.2

One should, from the beginning, encourage children to write the sign as

$$\sqrt{\ }$$

and not $\sqrt{\ }$, in order to give it maximum clarity, because, when handwritten, a mere tick easily degenerates further into something approaching / or the digit 1.†

Also notice that it is really quite absurd to write $\sqrt{12}$ as $\sqrt{\overline{12}}$ with an unnecessary bracket over the top, for that is what the vinculum (that horizontal line) is; just as absurd, in fact, as to write $12^2$ as $(12)^2$ with similarly unnecessary parentheses. With the square root of $12x$, of course, the use of brackets becomes *necessary*, $\sqrt{(12x)}$ or $\sqrt{\overline{12x}}$, exactly as with the square of $12x$, $(12x)^2$. On the other hand, $\sqrt{\dfrac{3h}{2g}}$ is unambiguous without any further embellishment; although, of course, if you are a solidus (/) fanatic, you will have to have recourse to ugly contrivances, like $\sqrt{[3h/(2g)]}$.

If you claim that $\sqrt{a} - b$ may be mistaken for $\sqrt{(a - b)}$, why are you not equally worried that $l - m^2$ may be mistaken for $(l - m)^2$, or $\ln 2 - \sqrt{3}$ for $\ln(2 - \sqrt{3})$, or $\Sigma_{r=1}^{n} u_r - 5$ for $\Sigma_{r=1}^{n}(u_r - 5)$? *If* you feel it essential to write $\sqrt{a} - b$ as $\sqrt{(a)} - b$ or $\sqrt{\overline{a}} - b$, should you not *also* insist on $l - (m)^2$, on $(\ln 2) - \sqrt{3}$ and on $(\Sigma_{r=1}^{n} u_r) - 5$?

Printers prefer $\sqrt{(x + y)}$ to $\sqrt{\overline{x + y}}$ for obvious technical reasons, just as they prefer $[p - (q - r)]^3$ to $(p - \overline{q - r})^3$, again avoiding a vinculum. For a similar reason, it is the printers who have forced us to adopt $(n - 1)!$ instead of $\lfloor n - 1$. Teachers *may* prefer to use $\sqrt{x + y}$ rather than $\sqrt{(x + y)}$ in writing, although the remarks in Section 3.7 about '$\sqrt{\ }$' as a functional prefix are particularly relevant and should be considered before a decision is taken.

But, please, let us have $\sqrt{t}$ and *not* $\sqrt{\overline{t}}$.

Since typographic problems are being discussed, and having just mentioned the solidus, this seems an appropriate place for a digression on the unsuitability of the solidus for use in teaching. If, since leaving school, you have got into the habit of writing your fractions in this way,

$$a/b \quad \text{instead of} \quad \frac{a}{b}, \qquad dy/dx \quad \text{instead of} \quad \frac{dy}{dx},$$

and so on, you would be most strongly advised to abandon the practice at once, before your students see you writing mathematics like this and start copying you. If you do not, the more thoughtless among your pupils will soon be writing

$$x + 1/y + 2$$

not only to denote

---

†The standard Linotype sign '$\sqrt{\ }$', as used by the printers of this book, will serve as a perfect example for teachers of the shape of square root sign *not* to be copied!

$$x + \frac{1}{y} + 2,$$

which is what it actually says, but also to mean

$$\frac{x+1}{y} + 2, \qquad x + \frac{1}{y+2} \quad \text{and} \quad \frac{x+1}{y+2},$$

because they will be too impatient or forgetful to write

$$(x + 1)/y + 2, \qquad x + 1/(y + 2) \quad \text{and} \quad (x + 1)/(y + 2)$$

respectively. Similarly, these characters will write both

$$\frac{1}{2\pi} \quad \text{and} \quad \frac{1}{2}\pi$$

in exactly the same way, as

$$1/2\pi,$$

and their written

$$\sin x/2$$

may mean either

$$\sin \tfrac{1}{2}x \quad \text{or} \quad \tfrac{1}{2}\sin x.$$

(Pupils always assume that those that have the misfortune to have to read their scripts will possess the psychic powers necessary to divine their meaning.) The ambiguities caused when the solidus is used by inexperienced or lazy practitioners are legion, and if you set a bad example by using it, you will have only yourself to blame. Moreover, your more perceptive colleagues will rightly be furious with you for undermining their careful attention to details like this. This is another of those undesirable habits that are easily picked up from lecturers in higher education, but are unsuitable for the schoolteacher to adopt.

It is most unfortunate that, for economic reasons, printers find themselves compelled to set so many fractions using a solidus. In an ideal world, this would never be done in any elementary textbook. In books (like this one) that will only be read by experienced adults, the use of the solidus has to be accepted, but the practice is only tolerable as long as readers are not tempted to copy such printed forms of fractions *in writing* when they are teaching children.

**3.3**

Returning now to square roots, however, genuine confusion is caused by lack of clear guidance as to the meaning that the sign '$\sqrt{}$' is actually given. Granted that the number 9 has two square roots, does the *symbol* $\sqrt{9}$ mean 3 or does it mean ±3? The writer has heard many students (and teachers!) strenuously affirm that they regard $\sqrt{9}$ as meaning ±3 and much further probing has been needed before they could be convinced that they did *not* use it in that way themselves; nor would it be advantageous for mathematics were it to be assigned such an interpretation.

When, for example, you draw the cartesian graph of $y = x^2 - 6x + 7$ and mark the points where it crosses the x-axis, you do *not* mark the points as in Fig. 3.1, regarding $\sqrt{2}$ as ±1·414 . . .; rather do you label the point at about $(4·414, 0)$ with the mark '$3 + \sqrt{2}$' and that at about $(1·586, 0)$ with the mark '$3 - \sqrt{2}$', as in Fig. 3.2, using $\sqrt{2}$ to mean the *positive* real number 1·414. . . . Similarly, do you not often write

$$\tan \frac{\pi}{3} = \sqrt{3}, \qquad \cos \frac{5\pi}{6} = -\frac{\sqrt{3}}{2}$$

and so on, and what does $\sqrt{3}$ mean there?

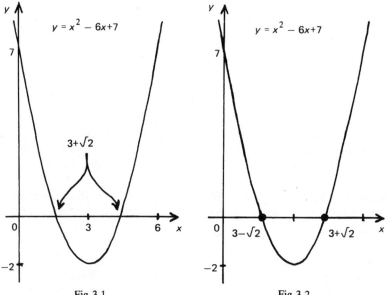

Fig 3.1                                    Fig 3.2

Also, if it really were the case that the symbol '$\sqrt{}$' contained within itself an ambiguity of sign, that favourite formula

$$\frac{-b \pm \sqrt{(b^2 - 4ac)}}{2a}$$

would not be written in that way at all, but as

$$\frac{-b + \sqrt{(b^2 - 4ac)}}{2a}.$$

Again, suppose a particle is projected vertically upwards with a speed $10 \text{ m s}^{-1}$ from the edge of a platform 20 metres above the ground. If the particle hits the ground after $t$ seconds, and the acceleration due to gravity is taken as $10 \text{ m s}^{-2}$, then $t$ satisfies

$$-20 = 10t - 5t^2,$$

that is,

$$t^2 - 2t - 4 = 0.$$

Hence $t = 1 \pm \sqrt{5}$ and, rejecting the negative value as inappropriate in this problem, the conclusion everyone writes down is that the particle hits the ground after $(1 + \sqrt{5})$ seconds, where again, by the symbol $\sqrt{5}$, we all mean a certain *positive* real number, $2 \cdot 236. \ldots$.

Instances like this of contexts in which the sign '$\sqrt{}$' is regularly employed make it quite clear that we use '$\sqrt{}$' to mean

'the non-negative square root of'.

Note also that if $\sqrt{5}$ *were* used to mean $\pm 2 \cdot 236 \ldots$, there would then be no convenient way of picking out the positive value; one would have to write $|\sqrt{5}|$ every time this was needed. It would be no good just writing $+ \sqrt{5}$ and hoping to create a distinction in this way, because that would mean that $+ \sqrt{5}$ was intended to be different from $\sqrt{5}$ and one does not want $+r$ to be different from $r$, whatever $r$ is. (What, moreover, would $4 + \sqrt{5}$ mean?) And would *you*, or anyone else, be prepared to write, for example,

$$\sin \frac{\pi}{3} = \frac{1}{2} |\sqrt{3}| \, ?$$

Used in the accepted way, however, there is no problem; with $x \geq 0$,

$$\sqrt{x} \text{ means the non-negative square root of } x.$$

If either square root of $x$ is admissible, $\pm\sqrt{x}$ is written, using an explicit ambiguous sign.

**3.4**

Observe that the modulus $|x|$ of a real number $x$ may be defined to be the 'absolute value' of $x$, that is,

$$|x| = x \text{ if } x \geq 0 \text{ and } |x| = -x \text{ if } x < 0.$$

Perhaps more succinctly, it may also be defined as

$$|x| = \sqrt{(x^2)},$$

that is, as[†] the non-negative square root of $x^2$.

This is in harmony with the definition of the modulus of a complex number $z = x + iy$ as

$$|z| = \sqrt{(x^2 + y^2)},$$

that is, as the non-negative square root of $(x^2 + y^2)$, and explains why, when the symbol $|-6|$ is used, it is not necessary to specify whether it means the 'absolute value' of the real number $-6$ or the modulus of the complex number $-6 + 0i$; the symbol $|-6|$ will have the same value, 6, with either interpretation.[‡] When choosing notations, mathematicians, as will be seen again in a moment, are not always as careful as this to preclude ambiguity. Let us at least be grateful when symbols do have this property!

Avoid, incidentally, the common error of saying that

$$\sqrt{(1 - \sin^2 t)}$$

is $\cos t$. It is not: it is $|\cos t|$.

---

[†]There is a definite reason for including the brackets in this definition. If $x$ is positive, they are unnecessary: for example, $\sqrt{9^2}$ is 9 whether it is calculated as $\sqrt{(9^2)} = \sqrt{81} = 9$ or as $(\sqrt{9})^2 = 3^2 = 9$. But, if $x$ is negative, the second method involves first considering the square root of a negative number; the use of the symbol $\sqrt{x}$ when $x$ is negative will come under critical scrutiny later.

[‡]Observe that both uses can be seen on the left side of the inequality

$$\left| |z_1| - |z_2| \right| \leq |z_1 + z_2|, \qquad z_1, z_2 \in \mathbf{C},$$

a statement that would be confusing were the two definitions not compatible.

This mistake, among others, is seen in the following argument, which purports to show that

$$\mathrm{ch}^{-1} \sec u = \mathrm{th}^{-1} \sin u.$$

$$\text{`ch}^{-1} \sec u = \ln[\sec u + \sqrt{(\sec^2 u - 1)}]$$

$$= \ln(\sec u + \tan u)$$

$$= \ln \frac{1 + \sin u}{\cos u}$$

$$= \frac{1}{2} \ln \frac{(1 + \sin u)^2}{1 - \sin^2 u}$$

$$= \frac{1}{2} \ln \frac{1 + \sin u}{1 - \sin u}$$

$$= \mathrm{th}^{-1} \sin u.\text{'}$$

The first step requires $\sec u \geqslant 1$ for $\mathrm{ch}^{-1} \sec u$ to be *defined*. [See Section 2.13.] Hence

$$(2k - \tfrac{1}{2})\pi < u < (2k + \tfrac{1}{2})\pi.$$

[The assumption that $u \neq (l + \tfrac{1}{2})\pi$ is also needed at other places in the argument.] Then, in the second step, $\sqrt{(\tan^2 u)}$ has been replaced by $\tan u$ instead of by $|\tan u|$. This involves the additional restriction that

$$m\pi \leqslant u < (m + \tfrac{1}{2})\pi.$$

The formula is therefore only proved (and is clearly only true) for

$$2n\pi \leqslant u < (2n + \tfrac{1}{2})\pi.$$

## 3.5

So far, $\sqrt{x}$ has only been used when $x \geqslant 0$. Is it a good idea to write $\sqrt{-1}$? Despite the fact that perhaps 4 out of 5 books do so, it is proposed to argue below that the answer should be an emphatic 'no'.

It cannot be too strongly or frequently emphasized to pupils that, unlike the real numbers, the complex numbers do *not* form an *ordered* system. One can say of two complex numbers $z_1$ and $z_2$ that either $z_1 = z_2$ or $z_1 \neq z_2$; unequal complex numbers can *not* be further classified by saying that '$z_1 > z_2$' or '$z_1 < z_2$', as one can for real numbers. The

ordering of **R** is the one important feature that does *not* carry over into
**C**.

Similarly, it makes perfect sense to describe one complex number $z_1$
as the negative of another number $z_2$ and to write $z_1 = -z_2$. But,
whereas the non-zero real numbers may be classified as 'positive' or
'negative' in an absolute sense because of the presence of an ordering
relation ($x > 0$ or $x < 0$), a similar classification for non-zero complex
numbers is impossible.

Indeed, to digress slightly, pupils' attention is not always drawn to
this distinction between these two contexts in which the word 'negative'
occurs. There is the absolute sense of 'is negative' applied to real
numbers; this is really a singulary relation (see Section 1.4). There is
also the usage in the binary relation 'is the negative of', which provides
the only legitimate occurrence of the word 'negative' in the study of
complex numbers. Other systems in which the second (binary) use is
possible, but not the first, include, of course, vectors and matrices.

Now there are two complex numbers, $i$ and $-i$, whose square is $-1$
and *each is the negative of the other*. But one can NOT say that $i$ is 'the
positive square root of $-1$' and $-i$ is 'the negative square root of $-1$';
*any such statement is in direct conflict with the structure of the complex
field*, **C**. This is why it seems wholly undesirable to write $i = \sqrt{-1}$ and
$-i = -\sqrt{-1}$. We have agreed that, if $x \geq 0$, $\sqrt{x}$ means the non-negative
square root of $x$. This selection of a distinguished member of the pair of
square roots is possible because of the ordering of the real field, **R**. In
**C**, where no such ordering is possible, the two square roots of a number
occur on an exactly equal footing and, in general, there is no
comparable procedure for selecting a distinguished member of the pair.
When $i$ is used[†] it never means one definite, eternally-labelled square
root of $-1$; it represents either one of the square roots and, whichever it
does, the other square root automatically becomes $-i$. It has the
property that whenever $i^2$ appears, it can be replaced by $-1$, but that is
*not* the same thing as saying that $i$ can be replaced by $\sqrt{-1}$.

Consider the following well-known type of fallacy,

$$-1 = i^2 = \sqrt{-1} \cdot \sqrt{-1} = \sqrt{(-1) \cdot (-1)} = \sqrt{1} = 1,$$

which the pupil is invited to expose. The usual explanation is that, when
a square root is taken, two choices can be made and if (by some
unspecified instinct) the appropriate choice is made on each occasion,
the fallacy will disappear. Of course, this 'explanation' will make the

---

[†]One must, of course, apologize to engineers who insist on using $i$ for current and so have
to use $j$ for the complex unit. But most mathematicians will continue to use $i$ in the way
they have always done and invent, if necessary, new symbols for their currents.

equation appear correct by taking suitable choices, but it scarcely provides a satisfactory resolution of the paradox.

The basis of the fallacy really lies in the very notation, $\sqrt{-1}$, used. This symbol contains within itself the seeds of these discords. It is self-contradictory; it suggests the possibility of picking one square root in a consistent way by some magic decision process, which, as said before, is just not possible, in general, when $\sqrt{x}$ is used with $x$ negative.

Note that no amount of ingenuity will create a similar plausible paradox out of

$$7 = (\sqrt{7})^2 = \sqrt{7} \cdot \sqrt{7} = \sqrt{7 \cdot 7} = \sqrt{49} = 7.$$

Yet 7 has two square roots, just as much as $-1$ does. When the symbol $\sqrt{x}$ is *not* used with $x$ negative, everything is consistent; the distinguished member *can* be selected. Putting it another way, the formulae

$$\sqrt{(ab)} = \sqrt{a} \cdot \sqrt{b} \quad \text{and} \quad \sqrt{\frac{a}{b}} = \frac{\sqrt{a}}{\sqrt{b}}$$

cannot be applied successfully when $a < 0$ or $b < 0$.

An apparent solution to the dilemma would be to use $\sqrt{x}$ differently when $x < 0$ from the way we have agreed that it is and should be used when $x \geq 0$, allowing it to represent the *pair* of square roots if $x < 0$ (so that one would have $\sqrt{(-a^2)} = \pm ia$). This was the usage that was so firmly rejected when $x \geq 0$. As you will perceive, however, even that would not dispose of the above type of fallacy: it would merely alter it to 'proving' that $\pm 1 = 1$.

Surely, a better solution—indeed, the only sensible one—is to decline ever to write $\sqrt{x}$ if $x < 0$. Then not only would there be no possible misunderstanding or conflict, but students might even understand complex numbers better.


### 3.6

There is, however, a price to pay, at least by those teachers who are addicted to solving quadratics by that ubiquitous formula. When solving a quadratic equation with complex roots, such as $z^2 + z + 1 = 0$, pupils who are encouraged to use 'the formula' will write

$$z = \tfrac{1}{2}(-1 \pm \sqrt{-3})$$

—a statement that will now be anathema to those who have been convinced by the above arguments—and then, perhaps, will transform this to

$$z = \tfrac{1}{2}(-1 \pm i\sqrt{3}).$$

The reason this slipshod method does not often lead to the sort of fallacy discussed earlier is that in most of these applications the two roots occur as a pair—contrast, for example, the distinguishable roots of the projectile equation mentioned earlier—so that writing $\pm i\sqrt{3}$ as $\pm\sqrt{-3}$ does not get the writer into the same sort of paradox trouble that writing $i\sqrt{3}$ as $\sqrt{-3}$ can so easily do.

Teachers who object to teaching pupils 'the formula' have no problem. They complete the square, as they would do with any other quadratic equation, getting

$$(z + \tfrac{1}{2})^2 = -\tfrac{3}{4}$$

and then obtain at once

$$z + \tfrac{1}{2} = \pm \tfrac{1}{2}i\sqrt{3}.$$

The price that formula fanatics would have to pay to be consistent would be to teach the formula as

$$\frac{-b \pm \sqrt{(b^2 - 4ac)}}{2a} \quad \text{if} \quad b^2 \geq 4ac$$

and

$$\frac{-b \pm i\sqrt{(4ac - b^2)}}{2a} \quad \text{if} \quad b^2 < 4ac.$$

It is understandable that this price is felt to be too high, which is presumably why $\sqrt{-3}$ continues to be seen, regardless of the mischief it causes.

### 3.7

But suppose that the stricter view being recommended is accepted as valuable. Then

$$y = \sqrt{x} \quad \text{is written only when} \quad x \geq 0 \quad \text{and} \quad y \geq 0.$$

Note that $\sqrt{}$ has now become a *function* (see Chapter 2), having as *both domain and range the set of non-negative real numbers*, S say, so that

$$S \twoheadrightarrow S : \sqrt{\phantom{x}}.$$

It is the possibility of picking out one of the square roots in a consistent fashion that makes $\sqrt{\phantom{x}}$ a function with the domain S chosen and the impossibility of doing so for $\sqrt{x}$ when $x$ is negative (or for $\sqrt{z}$ when $z$ is complex) that prevents it from being a function with domain **R** (or **C**). Thinking of $\sqrt{\phantom{x}}$ as a functional prefix should be beneficial to the whole presentation.

## 3.8

It has already been emphasized that, in complex algebra, the square roots of any complex number occur as a symmetric pair, each being the negative of the other. This is just a special case of the general result (de Moivre's theorem) that if $\zeta = \rho \operatorname{cis} \phi$ ($\rho > 0$), any number $w$ satisfying $w^n = \zeta^m$ (where $m$ and $n$ are coprime integers and $n > 0$) is of the form

$$\rho^{\frac{m}{n}} \operatorname{cis}\left(\frac{m}{n} \phi + \frac{2k\pi}{n}\right),$$

where $\operatorname{cis} \theta$ means $\cos \theta + i \sin \theta$, $k$ is an integer and $\rho^{m/n}$ denotes the *positive $n$th root of $\rho^m$*. (The qualification emphasized is relevant if $n$ is even.) It is readily seen that $w$ can take $n$ and only $n$ distinct values, which are given, for example, by $k = 0, 1, 2, \ldots, n - 1$. In the complex plane, the points represented are the vertices of a *regular $n$-sided* polygon inscribed in a circle, centre O, radius $\rho^{m/n}$.

There is no general way of picking out a single, distinguished individual from this aggregate: it is the *set* of $n$ numbers, with no element preferred above the rest, that is the interesting object of study. For this reason, it is usually agreed that the symbol

$\zeta^{m/n}$ shall denote *any* element of this set of $n$ complex numbers.

This means, in particular, that $\zeta^{m/n}$ is *not a function* of $\zeta$ (unless $n = 1$). Moreover, the *relation* $w = \zeta^{m/n}$ is exactly equivalent to $w^n = \zeta^m$.

At this stage, an experienced reader may be objecting that he thinks he *does* know a way of choosing an easily distinguished element from the set. Why not, he will be saying, (if $\zeta \neq 0$) select $\exp(m/n \ln \zeta)$? The theory of these functions exp and ln is considered in some detail in Chapter 8, and in Chapter 9 this problem of the notation for powers is examined further. But although it will then be found that in *certain* circumstances this procedure *will* pick out one element from the set in a consistent way, it will also become clear why it is *not* possible to do this in general in any useful fashion, substantiating the assertions above.

In fact, the use of $\zeta^{m/n}$ to represent any element of a set of $n$ numbers is not an entirely satisfactory symbolic choice, natural though it seems in this context. It is made clear in Chapter 9 that the symbol $b^a$ is inherently capable of two different interpretations and it is because mathematicians sometimes use it to mean one thing and sometimes the other that all these undesirable misunderstandings are generated.

**3.9**

Because of these differences in the use of $\sqrt{x}$ ($x \in S$, the set of non-negative real numbers) and $z^{1/2}$ ($z \in C$), there is indecision about the meaning of $x^{1/2}$ ($x \in S$), and similarly about $x^{m/n}$ when $n$ is even. It is a dilemma that cannot easily be resolved.

Sometimes $x^{1/2}$ is indeed used like $z^{1/2}$ to mean the pair of square roots of $x$. On the other hand, when required to integrate $\sqrt{x}$, the student is expected to say

$$\int \sqrt{x}\; dx = \int x^{1/2}\; dx = \tfrac{2}{3}x^{3/2} + A,$$

regarding $x^{1/2}$ as exactly the same as $\sqrt{x}$; and, if $|y| < 1$, the binomial series

$$1 - \frac{1}{2}y + \frac{1.3}{2.4}y^2 - \frac{1.3.5}{2.4.6}y^3 + \dots$$

has a sum $1/\sqrt{(1 + y)}$, which is usually written $(1 + y)^{-1/2}$.

What this means is that in some contexts $x^{1/2}$ with $x$ real and non-negative is being given a meaning different from that usually given to $(x + 0i)^{1/2}$ when the base is a complex number. In other words, when $n$ is even, $x^{m/n}$ is no longer *any* number $y$ satisfying $y^n = x^m$, as it would be when $x$ means the complex number $x + 0i$. When $x$ is real and non-negative, $x^{m/n}$ *may* stand for the specific positive $n$th root of $x^m$, as indeed it was required to do for $\rho^{m/n}$ earlier in this chapter. (Contrast the happier position noted previously in relation to the modulus, $|x|$ and $|x + 0i|$, where no ambiguity arose.)

Putting it another way, the cartesian graph of $y = \sqrt{x}$ is *certainly* that shown in Fig. 3.3, and is associated (in the way explained in Section 2.18) with a *function* with domain and range S. On the other hand, although on some occasions the graph of $y = x^{1/2}$ may also be taken to be as in Fig. 3.3, on other occasions, it may be taken to be merely that of the *relation* (equivalent to $y^2 = x$) exhibited in Fig. 3.4. You may like to try a similar exercise with the graph of $y = x^{3/2}$.

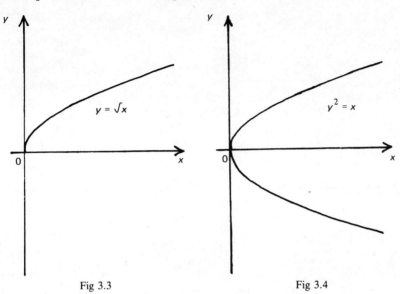

Fig 3.3                                    Fig 3.4

There is no way of reconciling these two interpretations without new notation: but don't let anyone tell you that mathematicians are consistent people!

# Chapter 4

# Infinity

---

## 4.1

However careful your presentation, there are always likely to be some pupils who have difficulty in seeing why division by zero is not possible. The only helpful stratagem, of course, is to look at numerous examples of the transition from $ax = b$ to $x = b/a$ for various $a$, $b$ with $a \neq 0$, and to contrast this with what happens in the case $a = 0$, when $0 . x = 0$ for all $x$. Thus the equation $0 . x = 1$ is seen to be inconsistent (that is, no $x$ is a solution) and the equation $0 . x = 0$ to be indeterminate (every $x$ is a solution).

But a genuine and serious problem arises as soon as your classes encounter that seductive symbol '$\infty$'. Not surprisingly, the pupil is anxious to know why he can *not* regard $\infty$ as the 'reciprocal of 0' and treat $\infty$ as just another number, boldy writing $1/0 = \infty$. There should be no astonishment that children take a long time to assimilate the concept and the correct use of the symbol: the work of many eminent mathematicians over many decades was needed to elucidate a consistent and acceptable way of handling the idea of 'infinity'.

You yourself may not be altogether clear whether writing $1/0 = \infty$ is such a heinous crime and in consequence your explanations may fail to satisfy your pupils. You may, after all, recall that with complex numbers that is more or less what mathematicians *do* write. Under the transformation $w = 1/z$, the image of the origin is taken to be the 'point at infinity'; the image of the number 0 is taken to be the 'number' $\infty$. It will, therefore, be useful to recall how this extension is justified in complex algebra. For your pupils, of course, these developments lie in the future, but understanding the course of events in complex algebra will help you to appreciate the point at issue in real algebra.

**4.2**

A sphere is drawn tangent to the complex plane at the origin O and I′ is the point of the sphere diametrally opposite O. A stereographic projection from I′ then maps every point P of the complex plane to a unique point P′ of the sphere, P′ being the other intersection of I′P with the sphere. Conversely, every point of the sphere except I′ is the image of a unique point in the plane (Fig. 4.1).

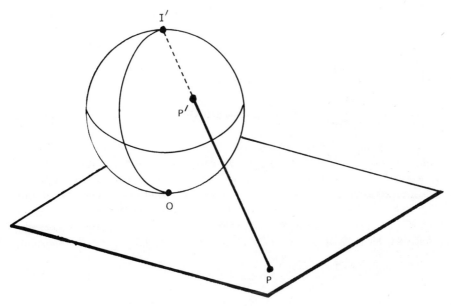

Fig 4.1

Now let W and Z be the points in the plane representing the numbers $w$ and $z$, and W′ and Z′ their stereographic images on the sphere under this projection. The transformation $w = 1/z$ is not a bijection [one-one correspondence] of the complex plane to itself, since the origin has no image. But as Z (and hence Z′) approaches O in any manner and W moves farther away from O in the plane along various paths, the point W′ always approaches the point I′ on the sphere. It is natural, therefore, to extend the complex plane by adjoining one extra 'ideal point' or 'point at infinity', regarding it as the missing partner I of I′ under the stereographic projection. The ideal point I then becomes the image of the origin O under the transformation $w = 1/z$, which is now a bijection of the *extended* complex plane onto itself. With I is associated an 'ideal number' $\infty$, which is the image of zero under the reciprocal transformation, and mathematicians write $w = \infty$ when $z = 0$.

Similarly, the general bilinear transformation

$$w = \frac{az + b}{cz + d} \quad (ad - bc \neq 0)$$

also becomes a bijection when both the domain and codomain are the *extended* complex plane. If $c \neq 0$, the image of $z = -d/c$ is $w = \infty$ and the image of $z = \infty$ is $w = a/c$. [If $c = 0$, then $z = \infty$ and $w = \infty$ correspond.]

### 4.3

The algebra in this extended system is entirely consistent. The new ideal number $\infty$ satisfies relations like

$$\infty + \zeta = \infty \quad \text{and (if } \zeta \neq 0) \quad \infty \cdot \zeta = \infty,$$

where $\zeta$ is an ordinary complex number. But this presents no difficulty: the symbol '$\infty$' can be incorporated into the complex number system without any conflict. It may have some peculiar extra properties, but no properties of the ordinary finite numbers are sacrificed. The transplant can take place without any rejection by the host system.

Care must, of course, be taken not to give a value to any 'indeterminate form', such as $\infty \cdot 0$, $\infty/\infty$ or (what may not be quite so obvious) $\infty + \infty$. [If

$$w_1 = 2 + z + \frac{1}{z}, \qquad w_2 = 5i - \frac{1}{z},$$

then, when $z = 0$, $w_1 = \infty$, $w_2 = \infty$ but, as $z \to 0$, $w_1 + w_2 \to 2 + 5i$.] But this caution is only a natural extension of the restraint one has to exercise in not writing $0/0$ or $0°$.

Why, then, does the same grafting process not succeed when the algebra is real? The trouble is our old friend, the ordering relation, which, as stressed in Section 3.5, evaporates in complex algebra but condenses again on to the structure as soon as the algebra becomes entirely real. Between any 2 real numbers $x$ and $y$, *one and only one* of the relations $x < y$, $x = y$, $x > y$ *must* hold. In particular, $x < x + 1$ for all $x$. If '$\infty$' could be added to the system and treated the same as any other symbol, we should have both $\infty < \infty + 1$ in accordance with this observation, but also $\infty = \infty + 1$ because of the special property of the symbol $\infty$. Conflicts like this are irresolvable; the very structure of an *ordered* field, like $\mathbf{Q}$ or $\mathbf{R}$, prevents the extension that is possible for $\mathbf{C}$.

**4.4**

For the young mathematician at school, use of the symbol '$\infty$' ought to be confined exclusively to the following contexts: (a) $\to \infty$, (b) $\to -\infty$, (c) $\sum^{\infty}$, (d) $\int^{\infty}$, (e) $\int_{-\infty}$. Each of these notations represents a highly stylized shorthand, and a whole swathe of conceptual preparation is necessary before any of these abbreviations is adopted by the student.

*Never* should the pupil be given occasion to write '$= \infty$' or to say that anything is 'equal to infinity'. The conventional usage above, $w = \infty$, relating to the ideal *complex* number will not arise until much later in his education.

There are, of course, strong temptations to reject this counsel of perfection with expressions like $\tan \frac{1}{2}\pi$, but it is much better resolutely to *avoid* any statement such as

$$\text{'}\tan \tfrac{1}{2}\pi = \infty\text{'}.$$

The minute saving of ink or chalk is no compensation for the harm such remarks may do when they are offered as substitutes for more precise and acceptable statements such as

$$\tan x \to \infty \quad \text{as} \quad x \nearrow \tfrac{1}{2}\pi.$$

[This notation is discussed in Section 5.1.]
    Notice also that, although

$$f(x) \to l \quad \text{as} \quad x \to a \quad \text{and} \quad f(x) \to l \quad \text{as} \quad x \to \infty$$

have alternative symbolizations as

$$\lim_{x \to a} f(x) = l \quad \text{and} \quad \lim_{x \to \infty} f(x) = l,$$

there are *no* corresponding substitutes for

$$f(x) \to \infty \quad \text{as} \quad x \to a \quad \text{and} \quad f(x) \to \infty \quad \text{as} \quad x \to \infty.$$

Any statement (in real algebra) that

$$\lim f(x) = l$$

implies that $l$ is a *real number*, never the artificial symbol '$\infty$' (or '$-\infty$').

By insisting on this practice, mathematicians are once again striving to avoid the phrase '$= \infty$' and thereby reinforcing the doctrine expounded above that '∞' can *not* be adjoined to **R** as an 'ideal' real number if the axioms of **R** are not to be sacrificed.

Having said this, it is only fair to point out that there is a related piece of theory where exceptions to this precept are permitted and it is possible that these may have been worrying you. If you have had considerable experience of real analysis at college, you will probably have come across the more advanced ideas of upper and lower limits ($\overline{\lim}$ and $\underline{\lim}$) and will know that here mathematicians *do* allow themselves the liberty of writing statements like '$\overline{\lim} f(x) = \infty$' (with a sign '='). But such statements, which are highly conventional, describe a more subtle phenomenon than the ordinary limiting process and there are good reasons for dispensing with the usual restraints. Such relaxation, however, does *not* extend to the ordinary limit and so this sort of conventional use of '$= \infty$' will *not* arise at school.

You may also have had the misfortune to study under lazy lecturers who have chosen to write (referring to series of non-negative terms)

$$\Sigma u_n = \infty \qquad \text{and} \qquad \Sigma v_n < \infty$$

to mean

$$\Sigma u_n \text{ diverges} \quad \text{and} \quad \Sigma v_n \text{ converges.}$$

Needless to say, you should *not* copy this slovenly practice with school pupils, for the obvious reason that the notation forcefully suggests that the symbol '∞' *can* denote some specific real 'number' (with a 'magnitude') and using it in that way will therefore sabotage all your efforts to convince them that this is not so.

For a similar reason, the use of '∞' in intervals, as in

$$x \in (a, \infty) \quad \text{and} \quad y \in (-\infty, b]$$

to mean

$$x > a \quad \text{and} \quad y \leq b,$$

is much better *avoided*, certainly in elementary work. If you use $(a, b)$ and $[a, b]$, do so *only* with $a \in \mathbf{R}, b \in \mathbf{R}$.

# Chapter 5

# Limits and convergence

---

## 5.1

When discussing limits and, in particular, when defining the derivative

$$f'(x) = \lim_{h \to 0} \frac{f(x + h) - f(x)}{h},$$

it is important for the student to recognize that $h$ must be allowed to tend to zero through negative values as well as through positive values if the existence of the limit is to be established. For example, if $f(x) = |x|$, $f'(0)$ does not exist, although $[f(h) - f(0)]/h$ tends to limits as $h$ tends to zero through both positive and negative values. There is, of course, no suggestion here that left and right derivatives should be mentioned at school: merely a wish that the two-sided nature of the ordinary limiting process should be appreciated by all pupils.

The point is that unless $g(x)$ tends to a certain number when $x \nearrow a$ ($x$ tends to $a$ from below) and to the *same* number when $x \searrow a$ ($x$ tends to $a$ from above), then the limit of $g(x)$ as $x \to a$ is not defined and the symbol $\lim_{x \to a} g(x)$ is not written. [The notation $x \nearrow a$ and $x \searrow a$ is so much more graphic and self-explanatory† than the rather weak alternative $x \to a -$ and $x \to a +$ that it must surely be preferred when teaching; anyway, it is the one that will be used here.]

## 5.2

Now let us ask some questions. Are there any objections to writing

(1) $\dfrac{\sin x}{x} \to 1$ as $x \to 0$, (2) $\text{th} \dfrac{1}{x} \to 1$ as $x \to 0$,

---

†It is due to John G. Leathem (1871–1926).

(3) $\dfrac{1}{x^2} \to \infty$ as $x \to 0$,          (4) $\dfrac{1}{x^3} \to \infty$ as $x \to 0$,

(5) $\ln x \to -\infty$ as $x \to 0$?

You may find it interesting to consider your responses to this challenge before reading further.

Certainly, there is no objection to statement (1) as long as it has been proved. The result $(\sin t)/t \to 1$ as $t \to 0$ is fundamental for obtaining the derivatives of the circular functions. For example, if $f(x) = \sin x$,

$$\frac{\sin(x + h) - \sin x}{h} = \frac{2 \cos(x + \tfrac{1}{2}h)\sin \tfrac{1}{2}h}{h} = \cos(x + \tfrac{1}{2}h) \cdot \frac{\sin \tfrac{1}{2}h}{\tfrac{1}{2}h}$$

and it is because $(\sin \tfrac{1}{2}h)/(\tfrac{1}{2}h) \to 1$ as $h \to 0$ that $f'(x) = \cos x$. You will doubtless remember how this lemma is proved.

A number $t$ is taken with $0 < t < \tfrac{1}{2}\pi$ and $t$ is represented as an angle as in Fig. 5.1. Then, since

$$\triangle OAP < \text{sector } OAP < \triangle OAT,$$

$$\tfrac{1}{2}a^2 \sin t < \qquad \tfrac{1}{2}a^2 t \qquad < \tfrac{1}{2}a^2 \tan t.$$

So, since $0 < t < \tfrac{1}{2}\pi$,

$$\cos t < \qquad \frac{\sin t}{t} \qquad < 1.$$

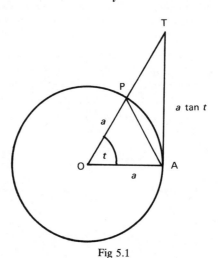

Fig 5.1

Hence

$$\frac{\sin t}{t} \to 1 \quad \text{as} \quad t \searrow 0.$$

Át least, this is what has *actually* been proved. Most textbook writers assume, without further comment, that what they have proved is

$$\lim_{t \to 0} \frac{\sin t}{t} = 1,$$

this being the result they need to use. But, of course, they have not established this; the geometric argument about magnitudes of areas only makes sense if $t > 0$. The omission is easily remedied. It merely requires the observation that, if $t < 0$,

$$\frac{\sin t}{t} = \frac{\sin(-t)}{-t},$$

and so

$$\frac{\sin t}{t} \to 1 \quad \text{as} \quad t \nearrow 0$$

also, giving the required result. But omissions like this do not help to emphasize the two-sided nature of the limiting process.

There is every objection to statement (2) above, since it is not true! This rather recondite example was merely introduced as counterpoint to (1) to persuade you that the above exhortations are not idle pedantry:

$$\text{th}\frac{1}{x} \to 1 \quad \text{as} \quad x \searrow 0, \quad \text{but} \quad \text{th}\frac{1}{x} \to -1 \quad \text{as} \quad x \nearrow 0; \quad \text{so} \quad \lim_{x \to 0} \text{th}\frac{1}{x}$$

is meaningless. [Incidentally, the cartesian graph of this function is worth sketching.]

Statements like (3) and (4) are often written, and probably (4) does not do too much harm. But, in the interests of accuracy, it should be appreciated that, while (3) resembles (1), $(1/x^2 \to \infty$ both as $x \nearrow 0$ and as $x \searrow 0)$, (4) is like (2), $(1/x^3 \to \infty$ as $x \searrow 0$, but $1/x^3 \to -\infty$ as $x \nearrow 0)$. A strong case can, therefore, be made out for avoiding assertions like (4). In this connexion, it is worth recalling the graphs of $y = x^{1/3}$ and $y = x^{2/3}$ and their gradients near the origin.

The last statement (5), however, is even more frequently written and is possibly much more damaging. Since $\ln x$ is defined only for $x > 0$,

$$\ln x \to -\infty \quad \text{as} \quad x \searrow 0$$

is the only claim that can legitimately be made. [Because of this feature of the logarithm, the graph of $y = 1/(\ln x)$ is interesting, with a stop-point at the origin.]

### 5.3

There is a fallacious argument, seen in some textbooks, that purports to show, for example, that

$$\lim_{x \to 0} x^x = 1.$$

'Let

$$u = x \ln x = \frac{\ln x}{\dfrac{1}{x}}.$$

Then, by l'Hôpital's rule (or some other reasoning),

$$\lim_{x \to 0} u = \lim_{x \to 0} \frac{\dfrac{1}{x}}{-\dfrac{1}{x^2}} = 0,$$

and so

$$\lim_{x \to 0} x^x = \lim_{x \to 0} e^u = 1.'$$

All that has *actually* been proved, of course, is that

$$\lim_{x \searrow 0} x^x = 1,$$

which is perfectly correct.

Indeed, no stronger result can be expected; because, when $x$ is a negative rational number with an even denominator, $x^x$ is not defined†
in real algebra. [Consider $(-3/4)^{-3/4}$.]

---

†There are also problems when $x$ is irrational and negative, but the above observation is adequate for the present purpose. See Chapter 9 for a discussion on the definitions of $b^a$.

So $\lim\limits_{x \nearrow 0} x^x$ is meaningless, and

$$\lim_{x \to 0} x^x \quad \text{does not exist.}$$

For a similar reason, not only does the *formula*

$$\frac{d}{dx}(x^x) = x^x(1 + \ln x)$$

only make sense if $x > 0$, but the very existence of a derivative for $x^x$ is impossible if $x \leq 0$.

It is worth examining the above argument in a little more detail. If $v$ is defined by

$$v = x \ln |x|,$$

then, as $x \to 0$, a (two-sided) limit for $v$ *does* exist and it is correct to claim that

$$\lim_{x \to 0} v = 0.$$

From this, the *valid* deduction that

$$\lim_{x \to 0} |x|^x = 1$$

can be made.

The graph of the continuous function $g$ defined by

$$g(x) = |x|^x \qquad (x \neq 0),$$
$$g(0) = 1,$$

is interesting and well worth your while investigating. If $x \neq 0$,

$$g'(x) = |x|^x(1 + \ln|x|),$$

while at $(0, 1)$ the graph of $y = g(x)$ is tangent to the $y$-axis.

## 5.4

This seems a good place to mention the cartesian graphs of two other functions whose behaviour may be unfamiliar to you. They are good examples to keep in your storeroom of useful curiosities, ready to be given an airing on appropriate occasions.

The first is that of

$$y = e^{\frac{1}{x}} \qquad (x \neq 0);$$

here $y \to 0$ as $x \nearrow 0$ and $y \to \infty$ as $x \searrow 0$.

The second is the author's favourite graph with a bounded discontinuity (that is, a point, $a$, at which $f$ is discontinuous but where $f(x)$ remains bounded as $x \nearrow a$ and as $x \searrow a$). It is the graph of $y = h(x)$, where

$$h(x) = \sqrt{\left(1 + \frac{1}{x} + \frac{1}{x^2}\right)} - \sqrt{\left(1 - \frac{1}{x} + \frac{1}{x^2}\right)} \qquad (x \neq 0).$$

It is not unlike the graph of $y = \text{th } 1/x$ mentioned earlier, but has the twin advantages of not requiring a hyperbolic function for its definition and of providing good practice in the careful manipulation of square root signs, as discussed in Chapter 3.

The first useful observation is that $h$ is an *odd* function $[h(-x) = -h(x)]$. But that does not mean, of course, that $h(0)$ ought to be defined as 0; one merely has to consider $x \mapsto 1/x$ to see that such a response is much too superficial. The dodge for determining the behaviour of $h(x)$ when $x$ is near zero is the slightly unusual one of rationalizing the *numerator*, giving

$$h(x) = \frac{\dfrac{2}{x}}{\sqrt{\left(1 + \dfrac{1}{x} + \dfrac{1}{x^2}\right)} + \sqrt{\left(1 - \dfrac{1}{x} + \dfrac{1}{x^2}\right)}}$$

Then, remembering the meaning of the square root sign [Chapter 3], $\sqrt{(x^2)}$ is $|x|$ and so

$$h(x) = \frac{|x|}{x} \cdot \frac{2}{\sqrt{(x^2 + x + 1)} + \sqrt{(x^2 - x + 1)}}.$$

Now

$$\frac{|x|}{x} = +1 \quad \text{if} \quad x > 0 \quad \text{and} \quad \frac{|x|}{x} = -1 \quad \text{if} \quad x < 0.$$

Hence

$$h(x) \to 1 \quad \text{as} \quad x \searrow 0 \quad \text{and} \quad h(x) \to -1 \quad \text{as} \quad x \nearrow 0.$$

The graph of $y = h(x)$ is shown in Fig. 5.2.

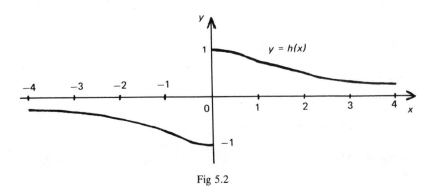

Fig 5.2

## 5.5

So far, consideration has only been given to the behaviour of $f(x)$ as $x$ tends to $a$. There is a corresponding point of interest about its possible behaviour as $x$ tends to infinity. This, again, concerns a distinction between one-sided and two-sided processes. To make the analogy clearer, some novel notation will be introduced, and it must be emphasized that this piece of ad hoc symbolism is non-standard.

Associate $x \nearrow \infty$ with $x \to \infty$ as conventionally written and $x \searrow \infty$ with $x \to -\infty$.† Consider the following four statements, the first two of which are partners for (1) and (2) above, while the others are simpler instances of similar behaviour.

(1a) $x \sin \dfrac{1}{x} \to 1$ both as $x \nearrow \infty$ and as $x \searrow \infty$.

(2a) th $x \to 1$ as $x \nearrow \infty$ and th $x \to -1$ as $x \searrow \infty$.

(1b) $\dfrac{3x + 2}{x} \to 3$ both as $x \nearrow \infty$ and as $x \searrow \infty$.

(2b) $\tan^{-1} x \to \tfrac{1}{2}\pi$ as $x \nearrow \infty$ and $\tan^{-1} x \to -\tfrac{1}{2}\pi$ as $x \searrow \infty$.

By exact analogy with $(\sin x)/x \to 1$ as $x \to 0$ and $3 + 2x \to 3$ as $x \to 0$,

†Note that writers who denote $x \nearrow a$ by $x \to a-$ and $x \searrow a$ by $x \to a+$ would, if they were consistent, denote $x \to +\infty$ by $x \to \infty-$ and $x \to -\infty$ by $x - \infty+$.

the statements in (1*a*) and (1*b*) might then be collapsed into

$$(1a) \quad x \sin \frac{1}{x} \to 1 \quad \text{as} \quad x \to \infty,$$

$$(1b) \quad \frac{3x + 2}{x} \to 3 \quad \text{as} \quad x \to \infty,$$

where '$\to \infty$' is now being used in a new, but more systematic, sense. On the other hand, (2*a*) and (2*b*) would continue to require *separate* statements for $x \nearrow \infty$ and for $x \searrow \infty$ (just like those for th $1/x$ and for $h(x)$ as $x \nearrow 0$ and as $x \searrow 0$).

The notations '$x \nearrow \infty$' and '$x \searrow \infty$' when $x$ is a real number† (and the new use of '$x \to \infty$') are *not* normal practice and so should not be used with pupils. They have been introduced here for the specific purpose of drawing attention to a useful analogy. Nevertheless they are notations that have certain advantages to commend them and, if they found favour, they might be worthy of consideration for teaching in the future. [On the other hand, however, it may be felt that ideas of tending to infinity 'from below' and 'from above' carry overtones of infinity being a real 'number' having a 'magnitude' and, if so, they might do more harm than good, for the reasons discussed in Chapter 4.]

### 5.6

It would be as well to prepare you for two mistakes in language, which you are almost certain to find your pupils making at times.

The first is the error of writing

$$\lim f(x) \to l.$$

The two permissible statements are

$$(1) \quad f(x) \to l \quad \text{as} \quad x \to a \quad \text{or} \quad f(x) \to l \quad \text{as} \quad x \to \infty$$

and

$$(2) \quad \lim_{x \to a} f(x) = l \quad \text{or} \quad \lim_{x \to \infty} f(x) = l.$$

---

†There is no case at all for not writing $n \to \infty$ when $n$ is a positive integer variable, or $z \to \infty$ when $z$ is complex.

These two notations cannot be mixed. [There are some further remarks about restrictions on these notations in Section 4.4.]

The second is the more serious solecism of making a statement such as '$f(x) \to l$' when $l$ is not independent of $x$. To write anything like

$$\tan x \to x \quad \text{as} \quad x \to 0 \quad \text{or} \quad \frac{4n^3 + 3n}{2n - 5} \to 2n^2 \quad \text{as} \quad n \to \infty$$

is taboo.

If you feel that the warnings in this section are too elementary to be necessary, may the author plead that he has seen teachers actually make these mistakes themselves!

### 5.7

The discussion will now turn to the subject of convergence. There is a conflict of terminology in this topic, particularly concerning the use of the word 'divergent', which it is not easy to resolve.

The general term of a series will be denoted here by $u_n$ if the terms are all real and by $w_n = u_n + i v_n$ ($u_n, v_n \in \mathbf{R}$) if they are complex, where the first term may correspond either to $n = 0$ or to $n = 1$ as convenient. The sum of the series as far as the term with suffix $n$ will be denoted by $s_n$ in all cases.

At any one time in the theory, one is usually concentrating on the properties of either

(1) a series of non-negative real terms ($u_n \geqslant 0$ for all $n$—or, at least, for all sufficiently large $n$), or
(2) a series of real terms ($u_n \in \mathbf{R}$), or
(3) a series of complex terms ($w_n \in \mathbf{C}$)

and, although one's interest is focused on rather different features in these 3 cases, it is regrettable that the nomenclature has been allowed to become inconsistent.

With a series of non-negative terms, there is, of course, a simple dichotomy in behaviour according to whether the monotone sequence $(s_n)$ is bounded or not. If $(s_n)$ is bounded, then the series $\Sigma u_n$ converges: if $(s_n)$ is unbounded, then $\Sigma u_n$ diverges. Life is simple.

With a series of general real terms, a more detailed 5-way classification emerges as valuable.

| *Property* | *Description* |
|---|---|
| 1. $\Sigma \lvert u_n \rvert$ converges | $\Sigma u_n$ AC ('absolutely converges' or 'is absolutely convergent') |

2. $\Sigma u_n$ converges but $\quad$ $\Sigma u_n$ CC
$\qquad$ $\Sigma \mid u_n \mid$ diverges $\qquad$ ('conditionally converges')
3. $s_n \to \infty$ or $s_n \to -\infty$ $\quad$ $\Sigma u_n$ D ('diverges')
4. $\Sigma u_n$ does not converge but $\quad$ $\Sigma u_n$ OB ('oscillates boundedly')
$\qquad$ $(s_n)$ is bounded
5. $\Sigma u_n$ does not diverge but $\quad$ $\Sigma u_n$ OU
$\qquad$ $(s_n)$ is unbounded $\qquad$ ('oscillates unboundedly')

Notice particularly that the abbreviations AC, CC, D, OB, OU are being used for the *verbs* indicated; that is, they denote specified *singulary relations* (see Section 1.4). Incidentally, the old-fashioned terms 'oscillates finitely' for OB and 'oscillates infinitely' for OU should certainly be avoided, as they introduce an unnecessary and misleading further use of words derived from 'infinity'.

A series $\Sigma w_n$ of complex terms could, in principle, be classified into one of 25 categories according to the 5 behaviour patterns possible for the x-axal part, $\Sigma u_n$, and the 5 for the y-axal part, $\Sigma v_n$. But this would not be a profitable exercise. Indeed, interest in $\Sigma w_n$ is usually confined to deciding which of the following 3 categories it falls into.

| *Property* | *Description* |
|---|---|
| 1. $\Sigma \mid w_n \mid$ converges | $\Sigma w_n$ AC |
| 2. $\Sigma u_n$ and $\Sigma v_n$ converge but $\Sigma \mid w_n \mid$ diverges | $\Sigma w_n$ CC |
| 3. $\Sigma w_n$ does not converge | |

And it is for this third, miscellaneous category (which includes 21 of the elementary behaviour patterns mentioned) that no satisfactory name exists. It is common to say that such a series 'diverges', but this is *not* a very sensible description. For one thing, it means that a series like $\Sigma(-1)^n n$ oscillates unboundedly (and does not diverge) when considered as a series of real terms, but diverges if considered as a series of complex terms. It would be nice to have a more distinctive designation of a non-convergent series, saying perhaps that it 'meanders' or 'wanders', but, until such a term is agreed on, a neutral and fairly acceptable (although not very elegant) description is 'non-converges'.

An alternative solution to the problem would be to *keep* the word 'divergent' just to mean 'non-convergent'—there would be no conflict in the case of series of non-negative terms—and to coin a new term, such as 'escaping' or 'receding' [although preferably *not* beginning with the letter 'R'!—see Section 2.9], to describe a series of real terms for which either $s_n \to \infty$ or $s_n \to -\infty$.[†] For the present, all one can do is to alert you to this further example of mathematicians' inefficiency.

---

†Just calling this behaviour *'proper divergence'* seems a particularly feeble solution.

The topic of series is taken up again in Chapter 13, where the particular problems of series *expansions* are discussed.

## 5.8

Going back a stage in the theory, from series to sequences, there is another decision that has to be taken and it is one on which, as you will discover, writers have differed. The word 'convergent' is correctly applied to suitable *series*, but should it ever be applied to *sequences*? Specifically, if $s_n \to l$ as $n \to \infty$, should one say that the sequence $(s_n)$ converges or merely that it tends to a limit? As many teachers have found, there are clear advantages in adopting the latter policy and keeping back the word 'converges' until series are studied. It is another way of reinforcing the difference in concept between a sequence and a series.

When the behaviour of a sequence is being investigated, the statement that the sequence $(s_n)$ tends to a limit (as well as that $s_n$ tends to a limit) is going to be made in any case. There is no additional insight to be gained by preempting another word 'converges' to say the same thing. It is with series, where the *sum* to $n$ terms tends to a limit and, in consequence, the *series* is said to converge, that the word is first needed. It is sound practice, therefore, to reserve the word for that purpose.

Similarly, since one wants to refer to the *terms* of a *series*, it is probably best to talk about the *members* (rather than the terms) of a *sequence*. Indeed, one can go further and suggest that the preferred terminology could well be

*term* of a *series*
*member* of a *sequence*
*element* of a *set*,

(which is the pattern followed in this book). But since there is, unfortunately, no very convenient alternative phrase for 'membership relation' in set language, the use of the word 'member' is always likely to be somewhat flexible.

These observations draw attention to another latent source of confusion. It is common to use braces (that is, curly brackets, { },) as part of the notation for sequences as well as for sets, writing sequences as $\{s_1, s_2, \ldots, s_n, \ldots\}$ or $\{s_n\}$. This custom can, however, sometimes lead to ambiguity and is not, therefore, very shrewd educational practice. As sequences, for example,

$$\{1, 2, 3, 4, \ldots\}, \quad \{2, 1, 4, 3, 6, 5, \ldots\}, \quad \{1, 1, 2, 2, 3, 3, \ldots\}$$

are all different: as sets, however (assuming the sequences continue in the obvious ways), they are all the same set, **P**.

There is a good case, therefore, for restricting the use of braces to sets and using parentheses (round brackets) for sequences: $(s_1, s_2, \ldots, s_n, \ldots)$ or $(s_n)$. This is the notation chosen for this book; it harmonizes with the notation for coordinates and other *ordered* sets of numbers $(x, y, z)$, $(\xi_1, \xi_2, \ldots, \xi_m)$, but may perhaps be criticized for not being sufficiently distinctive. It is also an awkward choice if one ever wants to write $s(n)$ instead of $s_n$, when the sequence becomes $(s(n))$.

At present, however, the use of braces for both concepts seldom causes serious difficulty and so it hardly seems necessary to urge you to abandon the notation $\{s_n\}$ for sequences if you find it attractive. But teachers should always take note of such potential causes of misunderstanding and be prepared to modify their habits with future classes once a problem has arisen or is anticipated.

# Chapter 6

# Circular and hyperbolic functions

---

## 6.1

The general procedure by means of which $\sin^{-1}$, $\mathrm{ch}^{-1}$ and so on are created as *functions* is described in Section 2.13. But there are some other observations worth making about more specific aspects of the circular and hyperbolic functions, direct and inverse, which are more elementary even than the remarks there, but would have interrupted the general discussion on functions.

From the point of view of serious mathematics, it is unfortunate, even though inevitable, that the circular functions first make their appearance as functions of an *angle* rather than as functions of a *real number*. [This slight misuse of language is discussed in Section 2.20.] Some students find it difficult to free themselves from this early attitude and all experienced teachers will have had the shock of seeing a pupil calculate something like $\int_0^{\sqrt{3}} dx/(1 + x^2)$ to be $60°$. The real variable $x$ in $\int dx/(1 + x^2)$ is, of course, exactly the same $x$ as that in $\int (1 + x^2)dx$ and there is no more justification for replacing $x$ by $(180x/\pi)°$ in the one case than in the other.

The true importance of the circular functions is simply that they are *periodic* functions of a real variable and, in a fairly precise mathematical sense, they are the *simplest* examples of such functions: think, for example, of how, using Fourier series, they provide the building blocks for constructing a very wide class of periodic functions.

At what stage does the transition from a function of an angle to a function of a real number take place? It must certainly occur *before* the derivative of any circular function is obtained. Results like, for example, $d/dx\,(\cos x) = -\sin x$ depend on the fact that

$$\lim_{t \to 0} \frac{\sin t}{t} = 1$$

and here $t$ is a real number. The usual proof of this result (mentioned for a different reason in Section 5.2) requires, however, when $t > 0$ that the number $t$ be represented as an angle (measured in 'radians') in order to use the formula $\frac{1}{2}a^2t$ for the area of the sector OAP (Fig. 5.1). The 'radian', of course, provides the bridge between the circular functions being functions of an angle and of a number. The radian is introduced as a unit of angle. But, if sound foundations for future work are to be laid, it is necessary to move fairly quickly forward, and to emphasize that *the circular functions have now become functions of a (real) number*, so that sin 1·27 is a well-defined number, just as is 1·27³.

The definition has made sin a function from **R** into **R** with range $[-1, 1]$; that is,

$$\mathbf{R} \rightarrow \mathbf{R} : \sin$$

or

$$\mathbf{R} \twoheadrightarrow [-1, 1] : \sin.$$

Similarly, there is a function

$$\mathbf{R} \setminus S \twoheadrightarrow \mathbf{R} : \tan,$$

where S is the set $\{(2n + 1) \, \pi/2 : n \in \mathbf{Z}\}$, and so on. Later, of course, these mappings will be extended to define functions of a complex variable.

The letter $x$ in the expression sin $x$ must be clearly perceived to be a number before the derivative is discussed. It is fundamental to the concept of a derivative, whether of $x^3$ or of sin $x$, that $x$ is a real number, not an angle. Failure to emphasize this is the cause of much confusion. There will be opportunities to stress it again when pupils meet the inverse functions, when they integrate expressions involving the circular functions and again when they expand these as power series.

### 6.2

Another very fruitful area for misunderstanding is the use of the inverse circular and hyperbolic functions in integration. It will be no novelty to a veteran campaigner to read solutions that say things like

$$\text{`}\int_0^1 \frac{\mathrm{d}x}{1 + x^2} = \frac{\pi}{4} + m\pi\text{'}$$

and

$$\int_0^{1/2} \frac{dx}{\sqrt{(1 - x^2)}} = \frac{\pi}{6}, \quad \text{or, more generally,} \quad n\pi + (-1)^n \frac{\pi}{6}.$$

But the fact that such mistakes occur at all may come as a surprise to the tiro.

Obviously, the most fundamental criticism of such answers is that a pupil who imagines that any ambiguity is *possible* in the value of the above integrals needs to revise rather carefully the meaning of a definite integral. But the real cause of these mistakes is lack of understanding of the correct definitions of the inverse functions, which is one reason why so much attention has been devoted to clarifying the theory in Chapter 2. Much of the blame for the perpetration of such errors must be laid at the feet of examiners like those who were responsible for the questions mentioned in Section 7.2.

There is one feature of the classroom presentation of this work on integration that the author has always found perplexing. All teachers expect (or at least hope) that as soon as they mention either of the mappings $x \mapsto 1/x$ or $x \mapsto \ln x$, the cartesian graph of the function (for an appropriate domain) will flash instantly into each pupil's mind. At least, they usually take active steps to promote such a reflex response. In their work on integration, teachers probably make nearly as much use of the mappings $x \mapsto 1/(1 + x^2)$ and $x \mapsto 1/\sqrt{(1 - x^2)}$, yet they seem to have no similar expectations of familiarity with the graphs involved there, and even the attention they give to the graphs of the functions occurring in their integrals, $x \mapsto \tan^{-1} x$ and $x \mapsto \sin^{-1} x$, may be quite perfunctory. Why are these pictorial aids so neglected? It is very puzzling.

It will be an instructive exercise to inspect the graphs (Fig. 6.1) of the 5 integrands and their integrals that tend to be brought together in this fragment of theory. For some readers it may be the first time they have ever looked carefully at the whole set. All the graphs are drawn to the same scale and it is a useful visual aid to plot the curves in this way on separate acetate sheets, so that pairs of graphs can be superposed using an overhead projector. The integrals are, of course, indeterminate to the extent of the usual arbitrary constant and the positions of their graphs relative to the $x$-axis have been chosen for simplicity. (In each of the lower diagrams in Fig. 6.1(d) and (e), two graphs, which are strictly speaking quite independent, have, perhaps imprudently, been plotted using the same axes.)

There are many noteworthy observations on these graphs. For example, the extra closeness of the curve $y = 1/(1 + x^2)$ to the $x$-axis,

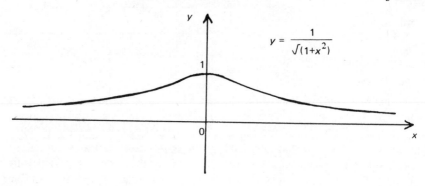

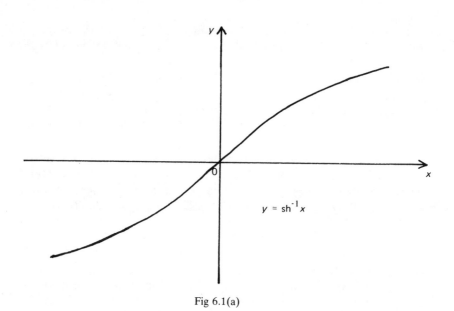

Fig 6.1(a)

compared with that of $y = 1/\sqrt{(1 + x^2)}$, is sufficient to make the integral over the whole domain convergent, as is seen by comparing the graphs of $y = \tan^{-1} x$ and $y = \text{sh}^{-1} x$ as $x \nearrow \infty$ and as $x \searrow \infty$. [See Section 5.5 for this notation.] A similar study of $y = 1/(1 - x^2)$ and $y = 1/\sqrt{(1 - x^2)}$ as $x \nearrow 1$ and as $x \searrow -1$ illuminates the asymptotic behaviour of the integral in the former case and the stop-point behaviour of the integral in the latter. It will also remind you that $\int_0^1 dx/\sqrt{(1 - x^2)}$ is *not* an ordinary

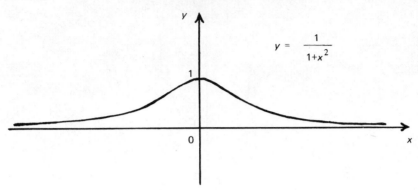

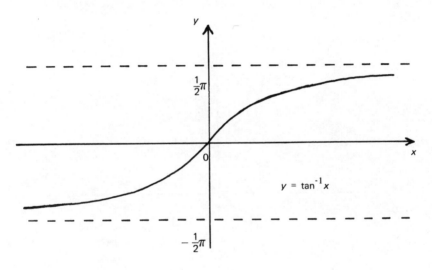

Fig 6.1(b)

Riemann integral, but a limit integral, and simply to write

$$\int_0^1 \frac{dx}{\sqrt{(1-x^2)}} = [\sin^{-1}x]_0^1 = \tfrac{1}{2}\pi - 0 = \tfrac{1}{2}\pi$$

is to suppress some very necessary discussion.

Juxtaposing pairs of graphs such as $y = 1/(1+x^2)$ and $y = \tan^{-1}x$

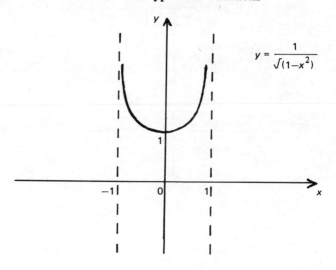

$$y = \frac{1}{\sqrt{(1-x^2)}}$$

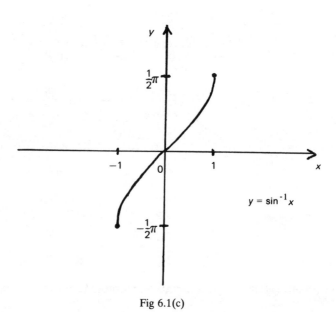

$$y = \sin^{-1} x$$

Fig 6.1(c)

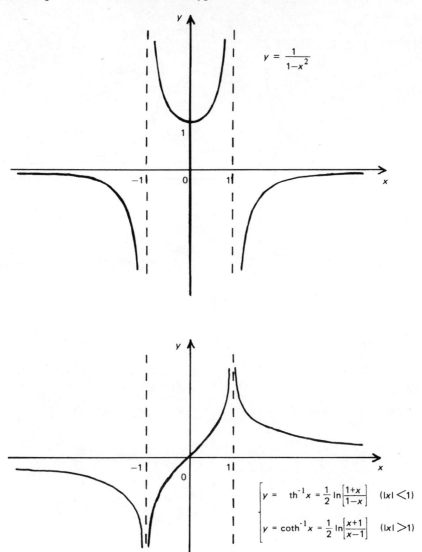

$$y = \frac{1}{1-x^2}$$

$$\begin{cases} y = \mathrm{th}^{-1}x = \dfrac{1}{2}\ln\left[\dfrac{1+x}{1-x}\right] & (|x| < 1) \\[3mm] y = \coth^{-1}x = \dfrac{1}{2}\ln\left[\dfrac{x+1}{x-1}\right] & (|x| > 1) \end{cases}$$

Fig 6.1(d)

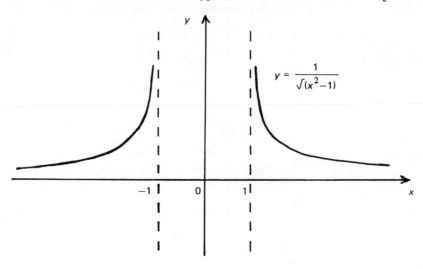

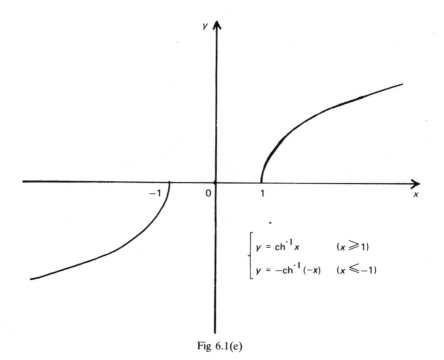

$$\begin{cases} y = \text{ch}^{-1} x & (x \geqslant 1) \\ y = -\text{ch}^{-1}(-x) & (x \leqslant -1) \end{cases}$$

Fig 6.1(e)

should help to avoid common mistakes in integration like

$$\text{`}\int_{-1}^{1} \frac{dx}{1 + x^2} = [\tan^{-1} x]_{-1}^{1} = \tfrac{1}{4}\pi - \tfrac{3}{4}\pi = -\tfrac{1}{2}\pi\text{ ,}\text{'}$$

as well as the errors previously noted.

Observe also that the formula for $\int dx/\sqrt{(x^2 - 1)}$ is $\mathrm{ch}^{-1} x + A$ *only* when $x > 1$; $\mathrm{ch}^{-1} x$ is not defined for $x < 1$ and, in particular, not for $x < -1$. The corresponding formula for the integral when $x < -1$ is almost invariably omitted from calculus books: it is $-\mathrm{ch}^{-1}(-x) + B$. [These two formulae cannot easily be combined into a single formula with modulus signs, as can the formulae for $\int dx/x$ with $x > 0$ and $x < 0$.]

Finally, notice particularly the different indefinite integrals for $\int dx/(1 - x^2)$ according as $|x| < 1$ or $|x| > 1$. Students often find difficulty in choosing the correct form of integral needed to obtain

$$\int_{1}^{2} \frac{dx}{9 - x^2} \quad \text{or} \quad \int_{4}^{5} \frac{dx}{9 - x^2},$$

although, since the necessity of deciding whether to select $\ln t$ or $\ln (-t)$ is an inescapable feature of *all* integrals involving logarithms, perhaps their dilemma here is merely an indication that they have never adequately discussed the fundamental issues involved. [See Chapter 8.]

## 6.3

When the inverse trigonometric functions were considered in Chapter 2, and again above, the notation exclusively adopted (which, incidentally, is due to the astronomer Sir John Herschel, son of Sir William), has been $\sin^{-1}$, etc. In your reading, you will almost certainly have come across an alternative notation 'arcsin' (particularly in American and some continental books) and may wonder what the relative advantages of the two notations are. It is proposed to argue below that *the arcsin notation has no merit whatsoever* and can, in fact, be considered positively harmful, as it gives the student a misleading idea of what is involved in the definition of an inverse function as it is now presented.

To explain this claim, it is necessary to look at the origin of the 'arc' notation to realize that it seriously conflicts with the modern approach. It was common up to the sixteenth century to treat the circular functions not even as functions of an angle but as functions of an *arc*

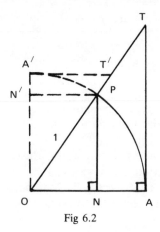

Fig 6.2

(this arc being less than a quadrant of a circle). This was achieved by always taking a circle of unit (or, at least, fixed) radius and (see Fig. 6.2) considering NP, AT, OT as the sine, tangent and secant† respectively of the *arc* AP. (The cosine, cotangent and cosecant of the arc AP are then the corresponding lengths N'P, A'T', OT' associated with a complementary arc such as A'P, where APA' is a quadrant of the unit circle.)

With this approach, the sine, tangent, etc. were thought of as *lengths* that were expressed as functions of another *length*, rather than as numbers expressed as functions of another number. One has only to point this out to realize what an advance it is to have our present treatment of these functions, which makes both $t$ and $\sin t$, etc. *ratios* of lengths and hence dimensionless numbers. When the earlier point of view was in vogue, it was, of course, entirely reasonable to use 'arcsin $u$' to mean the 'arc whose sine is $u$', that is, the length of the arc AP of the unit (or standard) circle for which the length of NP is $u$.

The shift away from this concentration on the arc began in the work of Rheticus (1514–1576) and culminated in that of Euler two centuries later. Of course, even with the modern definition of $\sin t$ as a *ratio*,

---

†The names 'tangent', to mean a certain segment of a 'touching line', and 'secant', for a segment of a 'cutting line', which both date from the late sixteenth century, are, of course, easily explicable in terms of this diagram. The etymology of the word 'sine', however, is confused but interesting. The Latin word 'sinus' means 'bay' (in the geographic sense) or 'bosom' and was used in the twelfth century to translate the Arabic word 'jaib', which has similar meanings. Unfortunately, the assumed form 'jaib' of the word (written, without vowels, as 'jb') was a misreading by the translator. The intended Arabic word had been 'jiba', which had been used to transliterate the Hindu word 'jiva' or 'jya' meaning 'chord'.

See, for example, C. B. Boyer *A history of mathematics* (Wiley) page 278 or D. E. Smith *History of mathematics* Volume 2 (Dover) page 616.

NP:OP, it is not to be denied that, when OP = 1, sin $t$ is equal to NP and so can be *thought* of as a length, if necessary. The point is that it gives completely the wrong emphasis to the concept to *define* sin $t$ as a length. It is similarly wrong to treat the *inverse* sine as a length, and yet that is exactly what the prefix 'arcsin' predisposes us to do. (When $z$ becomes complex, to write $\sin^{-1} z$ as arcsin $z$ becomes even less justifiable.)

When a writer with an 'arc' obsession comes to the inverse hyperbolic functions, he has a problem. He may choose to press the analogy with the circular functions, writing arcsh, arcth, etc., even though these functions do not have the same significance for the geometry of the rectangular hyperbola ($x^2 - y^2 = 1$) as arcsin, arctan did for the geometry of the circle ($x^2 + y^2 = 1$). He is, however, more likely to introduce a new piece of mumbo jumbo, argsh and argth, for his inverse hyperbolic functions.

What the inverse sine of $u$ means today is a particular *number* whose sine is $u$ and similarly for all the other inverse circular and hyperbolic functions. Any notation that suggests otherwise is misleading. [One might not object so strongly to 'invsin' as one does to 'arcsin'—pocket calculator manufacturers, please note—but, even so, that is still *greatly* inferior to 'sin⁻¹'.]

By choosing the notations $\sin^{-1}$, $\tan^{-1}$, $\text{sh}^{-1}$, $\text{th}^{-1}$, etc. for the inverse functions, not only can you be absolutely systematic but, what is even more important, you also help to reinforce a concept that your students are nowadays introduced to (in U.S.A. as well as in U.K.!) quite early anyway. That is the general idea that, if (and only if) the function

$$D \rightarrow C : f$$

is bijective, an inverse function

$$C \rightarrow D : f^{-1}$$

can be defined, and this function one positively *wishes* to write $f^{-1}$, and *not* arg $f$ or anything silly like that.

As explained in Section 2.13, by choosing suitable domains for the circular functions, they can be arranged to be bijective and then, inverse to the functions

$$[-\tfrac{1}{2}\pi, \tfrac{1}{2}\pi] \twoheadrightarrow [-1, 1] : \sin \quad \text{and} \quad (-\tfrac{1}{2}\pi, \tfrac{1}{2}\pi) \twoheadrightarrow \mathbf{R} : \tan,$$

are

$$[-1, 1] \twoheadrightarrow [-\tfrac{1}{2}\pi, \tfrac{1}{2}\pi] : \sin^{-1} \quad \text{and} \quad \mathbf{R} \twoheadrightarrow (-\tfrac{1}{2}\pi, \tfrac{1}{2}\pi) : \tan^{-1},$$

and so on. With this notation, these procedures are clearly seen to be merely special instances of the general method of introducing a function $f^{-1}$ inverse to $f$. The relegation of the 'arc' notation to the status of a historic curiosity is long overdue.

### 6.4

An objection raised by some people who have not thought at all deeply about this is to say: 'My pupils are expected to write $\sin^2 x$ for $(\sin x)^2$, so how can they be expected to write $\sin^{-1} x$ for anything other than $(\sin x)^{-1}$?' Apart from the fact that there is already an accepted, unambiguous notation for $(\sin x)^{-1}$, namely $\csc x$, the teacher who says this is tilting at the wrong windmill. The notation $\sin^{-1}$ for the inverse sine function is completely logical: the notation $\sin^2$ as it occurs in $\sin^2 x$ is absolutely absurd and indefensible.

Were it not already sanctioned by long tradition, nobody today would advocate the introduction into school mathematics of such an idiotic notation as $\sin^2 x$ for $(\sin x)^2$, no matter how many sets of brackets it might save, so fundamentally does it conflict with the idea of a function being a mapping. Why? Well, the only *logical* meaning to give the symbol $\sin^2 x$ is $\sin \sin x$ [that is, $\sin(\sin x)$], *not* $(\sin x)^2$. Gauss saw this perfectly clearly and fulminated against such a maladroit symbolism: '$\sin^2 \phi$ is odious to me, even though Laplace made use of it; should it be feared that $\sin \phi^2$ might become ambiguous, which would perhaps never occur, or at most very rarely when speaking of $\sin(\phi^2)$, well then, let us write $(\sin \phi)^2$, but not $\sin^2 \phi$, which by analogy should signify $\sin(\sin \phi)$'.† But, alas, even his patriarchal authority was not sufficient to prevent it becoming established—and it is insidiously convenient.

There is a similar problem with the logarithm function. But whereas with the sine function, $(\sin x)^2$ is common and $\sin \sin x$ extremely rare, it is the opposite with the logarithm, where $\ln \ln x$ occurs much more often than $(\ln x)^2$. So, were it not for the baleful influence of that traditional $\sin^2 x$, one could, legitimately and happily, use $\ln^2 x$ with its obvious meaning of $\ln \ln x$. As it is, however, one is virtually precluded from using $\ln^2 x$, because of the possibility of confusion: a case of guilt by association.

It is now much too late to consider abandoning the conventional notation for the powers of the circular and hyperbolic functions, but at least let us be clear which notation is the offending one and not use the argument of one illogical notation ($\sin^2$, $\mathrm{sh}^2$) to encourage us

---

†Quoted (no. 1886) in R. E. Moritz *On mathematics and mathematicians* (Dover).

to use another bad one (arcsin, argsh), instead of the sensible one ($\sin^{-1}$, $\text{sh}^{-1}$).

It is best, however, always to *read* '$\sin^{-1}$' as 'sine minus one' [or as 'sine inverse'], *never* as 'sine *to the* minus one'; just as one would read '$f^{-1}$' as '$f$ minus one' or '$f$ inverse'.

Note that, given a bijective function $D \to C : f$ with inverse $C \to D : f^{-1}$, it makes sense in the case when $C = D$ to define the function $D \to D : f^{-2}$, which is the same whether $f^{-2}$ is interpreted as $(f^{-1})^2$ or as $(f^2)^{-1}$. But, because of the problems with $\sin^2$, one must never write $\sin^{-2} x$ (as one might, for example, be tempted to do when simplifying the integrating factor $e^{\int -2 \cot x \, dx}$).

Having explained all this, however, it is perhaps only honest to tell you, the teacher, that mathematicians *do* sometimes use $f^2(x)$ to mean $[f(x)]^2$. This is never done in analysis, but when functions are operated on as elements in algebraic structures, it may happen that, in the same piece of work, multiplication of functions, $x \mapsto f(x)g(x)$, and composition of functions, $x \mapsto f[g(x)]$, are both, in appropriate circumstances, significant operations. Mathematicians then choose one of the following representations: either

$fg : x \mapsto f(x)g(x)$ for multiplication    and

$$f{\circ}g : x \mapsto f[g(x)] \text{ for composition,}$$

or

$fg : x \mapsto f[g(x)]$ for composition    and

$$f . g : x \mapsto f(x)g(x) \text{ for multiplication.}$$

The two uses of $fg$ when $f = g$ correspond to the two possible meanings of $\ln^2$. But there is no reason whatever to confuse pupils by telling *them* this; the number who will be likely ever to represent both operations in the same piece of mathematics is negligible—and they are not the ones who will have difficulty in coping.

Chapter 7

# Inverse relations, arguments and polar coordinates

***

## 7.1

It was mentioned in Section 2.3 that some authors have used the notation $y = \text{Sin}^{-1} x$, with an upper case S, as an alternative way of expressing the relation $\sin y = x$, with a similar convention regarding other circular and hyperbolic relations. At the time, we were criticizing the claim that this defined a 'function', $\text{Sin}^{-1}$, whose 'principal value' was $\sin^{-1}$, pointing out that this was totally at variance with the modern definition of functionality. But this writer would like to put in a strong plea for the retention of the *notation*. Provided it is clearly understood that it is used to represent a *relation* (and does *not* define a function $x \mapsto y$), the notation $y = \text{Sin}^{-1} x$ seems to be a most valuable addition to our repertory. When working with the relation $x = \sin y$, it may be convenient to write it in a form where $y$ is the subject (just as it is sometimes useful to write the relation $x = y^2$ as $y = \pm \sqrt{x}$). Since it has been agreed that $\sin^{-1}$ is to be used for a function (with a certain domain and codomain), the notation $y = \text{Sin}^{-1} x$ is available and seems an estimable choice.[†] The large letter S suggests the 'large' number of values $y$ can take for a given $x \in [-1, 1]$, reminding the reader that the relation between $x$ and $y$ is one-*many*. There should be no confusion once it is taken as a rule that, in elementary work, all literal prefixes for *functions* (including exp, ln, lg, as well as the symbols for direct and inverse circular and hyperbolic functions) start with a *lower case* letter.[‡]

***

[†] It should, however, be noticed that this notation for these inverse circular and hyperbolic relations is non-standard, being of the form '$y = r(x)$' rather than '$x \, \rho \, y$'. It is a type of notation best kept just for these particular relations, where it happens to be especially convenient.

[‡] In advanced work, some non-elementary functions, such as Si, Ci, Ei, *are* conventionally written with capital letters, but no such functions arise at school.

Obviously, this piece of symbolism is not indispensable and some teachers may prefer to ignore it entirely. Nevertheless it is available if required and it can be very useful. For example, it may be appropriate at times to write a statement like

$$\mathrm{Cos}^{-1} \frac{\sqrt{5}-1}{4} = 2n\pi \pm \tfrac{2}{5}\pi$$

or to say that, if

$$\cos 2x = \sin x,$$

then

$$x = \mathrm{Sin}^{-1}\tfrac{1}{2} \quad \text{or} \quad \mathrm{Sin}^{-1}(-1).†$$

## 7.2

The case in favour of adopting a notation that is capable of representing both ideas—the function and the relation—and discriminating between them can be forcefully illustrated by the following sequence of questions taken from some S-level papers of a well-known Examination Board (which shall be nameless). All the examples are from corresponding papers of the same examination over a 13-year period.

---

†Although not, in fact, adopted in this book, an even *better* notation would be to use $\mathrm{Sin}^{-1} x$ to denote the *set* of numbers $y$ such that

$$x = \sin y,$$

so that the inverse of this relation would be written

$$y \in \mathrm{Sin}^{-1} x,$$

rather than $y = \mathrm{Sin}^{-1} x$. Thus, for example,

$$\mathrm{Tan}^{-1} \sqrt{3}$$

would be defined as the *set*

$$\{(m + \tfrac{1}{3})\pi : m \in \mathbf{Z}\},$$

and the solution of the above equation would be expressed as

$$x \in \mathrm{Sin}^{-1}\tfrac{1}{2} \cup \mathrm{Sin}^{-1}(-1).$$

**1948** Hence show that

$$\tan^{-1}\sqrt{\frac{a(a+b+c)}{bc}} + \tan^{-1}\sqrt{\frac{b(a+b+c)}{ca}} + \tan^{-1}\sqrt{\frac{c(a+b+c)}{ab}} = n\pi,$$

where $n$ is an integer, positive or negative, and the positive values of the square roots are taken.

**1949** Prove that

$$\tan^{-1}\tfrac{4}{3} + \tan^{-1}\tfrac{12}{5} = \pi - \sin^{-1}\tfrac{56}{65}.$$

**1951** Solve the equation

$$\tan^{-1}(x+1) + \tan^{-1}(x-1) = \tan^{-1}\tfrac{4}{7}.$$

**1955** Prove that one of the values of $\tan^{-1}\tfrac{1}{7} + 2\tan^{-1}\tfrac{1}{3}$ is $45°$.

**1956** Prove, without using tables, that one of the values of

$$2\tan^{-1}\tfrac{1}{2} - \tan^{-1}\tfrac{1}{7} \text{ is } \tfrac{1}{4}\pi.$$

**1957** Prove, without using tables, that

$$\tan^{-1}\tfrac{1}{2} + \tan^{-1}\tfrac{2}{3} + \tan^{-1}\tfrac{4}{7} = \tfrac{1}{2}\pi.$$

**1960** Solve the equation

$$2\sin^{-1}x + \sin^{-1}x^2 = \tfrac{1}{2}\pi.$$

In 1948, the occurrence of $n\pi$ in the question made it clear that the examiner that year was using $\tan^{-1}$ *relationally* (that is, to mean Tan$^{-1}$). [Also, curiously, he seems to have rejected the possibility of $n$ being zero, although that's beside the point!] Equally obviously, in the following year, $\tan^{-1}$ and $\sin^{-1}$ were being used for the *functions*. The question in 1951 was a candidate's nightmare. If it was set by the 1948 examiner or someone of his persuasion, he will have intended his equation to mean that 'Some value of Tan$^{-1}$ $(x+1)$, added to some value of Tan$^{-1}$ $(x-1)$, is equal to some value of Tan$^{-1}$ 4/7 and will have expected the students to discover *two* solutions, $x = -4$ and $x = \frac{1}{2}$. If, on the other hand, the equation was the responsibility of the 1949 examiner, he will have rejected the possibility of $x = -4$ as incompatible with *his* (and, one hopes, with *your*) meaning of 'tan$^{-1}$'

and will not have been satisfied unless the candidates gave an answer containing only *one* solution, $x = \frac{1}{2}$.

Overlooking the astonishing aberration of '45°' in 1955, it is clear from the phrases 'one of the values' that, in both 1955 and 1956, the relational use was holding sway, whereas, by 1957, $\tan^{-1}$ had again become a function. The equation of 1960 repeated the dilemma of 1951: a well-taught student who used $\sin^{-1}$ for the function will have obtained only *one* solution, $x = \frac{1}{3}\sqrt{3}$, but if he suspected that the turn of the 1956 examiner had come round again, he will have done better to humour him by producing *two* solutions, $x = \pm\frac{1}{3}\sqrt{3}$.

Even though this sort of confusion is partly due to the persistence of out-of-date ideas on functionality, it does show clearly that there is need for a precise and sensitive notation. The whole topic would be immeasurably clarified by the adoption of a distinctive symbolization for these inverse relations. Furthermore, the above line of reasoning yields a valuable bonus when it comes to complex algebra.

**7.3**

The argument of a non-zero complex number $z$ is only determined to within an integer multiple of $2\pi$ and on many occasions the particular value of the argument taken is immaterial. If $z = r$ cis $\theta$, $r \neq 0$, it is then appropriate to write

$$\theta = \text{Arg}\, z,$$

with an upper case A. For other purposes, as you will appreciate, it is more convenient to choose a precise, well-defined value of the argument; in other words, to make the argument of $z$ a *function* of $z$ for all $z \neq 0$. In keeping with the convention, this special value can in constrast be denoted by $\arg z$. For various technical reasons that are touched on in Section 8.7, the value chosen is the one that makes

$$-\pi < \arg z \leqslant \pi.$$

This fixes $\arg z$ as the numerically least value of $\text{Arg}\, z$, the positive choice being made in the only case where that rule is ambiguous. In other words, the definition being made is of a *function*

$$\mathbf{C} \setminus \{0\} \to \mathbf{R} : \arg$$

with range $(-\pi, \pi]$; or, if you like,

$$\mathbf{C} \setminus \{0\} \twoheadrightarrow (-\pi, \pi] : \arg.$$

Teachers often find it helpful to write

$$\text{Arg } z_1 + \text{Arg } z_2 = \text{Arg } (z_1 z_2)$$

to mean that, if any value of $\text{Arg } z_1$ is added to any value of $\text{Arg } z_2$, then the sum is *some* value of $\text{Arg } (z_1 z_2)$. (Other teachers prefer to avoid this type of statement.) It is, of course, necessary to emphasize, in contrast, that

$$\arg z_1 + \arg z_2 \quad \text{is not necessarily equal to} \quad \arg (z_1 z_2).†$$

(If $\arg z_1 = \arg z_2 = \frac{3}{4}\pi$, $\arg (z_1 z_2) = -\frac{1}{2}\pi$.)

These remarks about the argument are taken up in Section 8.6 in connexion with the logarithm function. But there is another different point about the argument that is worth noting.

**7.4**

It is quite *incorrect* to say that, if $z = x + i y$,

$$\arg z = \tan^{-1}\frac{y}{x},$$

as many students do (and not a few writers, who should know better). Nor is it correct to say that

$$\text{Arg } z = \text{Tan}^{-1}\frac{y}{x}.$$

Fixing the value of any *one* trigonometric expression involving $\theta$ (be it $\sin \theta$, $\cos \theta$ or $\tan \theta$) does *not* fix $\theta$, even to within a multiple of $2\pi$; there are always *two* possible radii vectores when the value of only one such expression is specified.‡ For example,

$$\arg(-1 - i) = -\tfrac{3}{4}\pi,$$

---

†Just as

$$\tan^{-1} a + \tan^{-1} b \text{ is not necessarily equal to } \tan^{-1} \frac{a + b}{1 - ab}.$$

‡Yes, all right; they *do* coincide sometimes, as when $\sin \theta = -1$; but that is irrelevant.

but

$$\tan^{-1}\frac{y}{x} = \tan^{-1} 1 = \tfrac{1}{4}\pi.$$

Similarly,

$$\text{Arg }(1 + i) = 2m\pi + \tfrac{1}{4}\pi \quad \text{and} \quad \text{Arg }(-1 - i) = (2n + 1)\pi + \tfrac{1}{4}\pi,$$

but, in *both* cases,

$$\text{Tan}^{-1}\frac{y}{x} = \text{Tan}^{-1} 1 = k\pi + \tfrac{1}{4}\pi,$$

where $k$ may be *any* integer, even or odd.

It is *essential* to say that Arg $z = \theta$, where

$$\cos \theta : \sin \theta : 1 = x : y : \sqrt{(x^2 + y^2)},$$

or something equivalent to this, so that the values of *two* trigonometric expressions are fixed, in order to give all of them the correct signs for the quadrant as well as the correct magnitudes.

## 7.5

There is a natural tendency to describe $r = |z|$ and $\theta = \text{Arg } z$ as polar coordinates in the complex plane. This certainly need not be discouraged, but it is worth pointing out to you as a teacher that there *is* a slight problem in saying that $(r, \theta)$ are plane polar coordinates.

When $r$ is the modulus of a complex number, it is required inexorably to be non-negative, as it is also when formulating the elementary definitions of the circular functions of a general angle. It is *possible* to adopt the same convention when $r$ is a polar coordinate and there are distinct advantages of consistency in doing so. Unfortunately, however, this convention can lead to slight complications at times and so some mathematicians are prepared once again to sacrifice consistency on the altar of expediency and to allow $r$, when it is a polar coordinate, to take *all* real values, including negative ones. This has the effect of making $(-r, \theta)$ the same point as $(r, \theta + \pi)$.

In any particular investigation, however, the convention that is being used with the polar coordinate $r$ is not a matter of indifference. For example, with $r \geqslant 0$, the rose curve $r = a \sin 2\theta$ has two loops ($\alpha$ and $\gamma$ in Fig. 7.1); with $r$ unrestricted, it has four. To obtain the four-loop

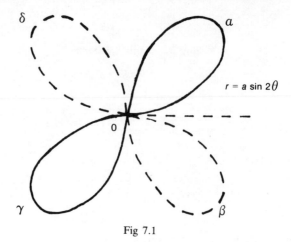

Fig 7.1

curve with the $r \geq 0$ convention, the graph drawn must be that of the relation $r^2 = a^2 \sin^2 2\theta$.

## 7.6

When the writer was at school, he remembers being puzzled why polar coordinates were taken in the order $(r, \theta)$, instead of in the order $(\theta, r)$, which at that time seemed to him more logical. After all, with cartesian coordiantes $(x, y)$, one always mentioned first the coordinate $(x)$ that was usually the independent variable in analysis, so why, with polar coordinates, did one mention that coordinate $(\theta)$ second? Only later did the reason become apparent. If they are taken in the order $(\theta, r)$, the new coordinates form a *left*-handed system. (See Fig. 7.2.) The usual order $(r, \theta)$ is taken to make the system right-handed, like $(x, y)$; that is, so that the jacobian $\partial(r, \theta)/\partial(x, y)$ shall be positive.

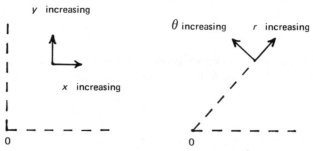

Fig 7.2

Incidentally, it is worth making clear to your pupils that the distinction between right-handed and left-handed coordinate systems is not just a feature of 3-dimensional euclidean space. It is present in any number of dimensions and is associated with the fact that the determinant of an orthogonal matrix, whatever its order, can only be either $+1$ or $-1$. It is just as impossible to superpose

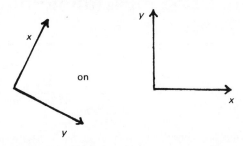

by a direct isometry *in the plane* (that is, without letting the framework rotate about an axis that takes it *out* of the plane and then back again), as it is to superpose

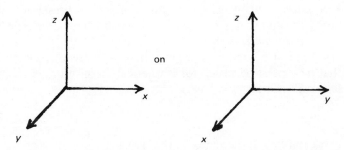

without performing a similar rotation that takes the framework into a 4-dimensional euclidean space before returning it to 3-space. Students often seem to have acquired the impression that the distinction between dextro and laevo systems (to use the scientists' terminology) is a specifically 3-dimensional phenomenon.

Chapter **8**

# Exponential and logarithmic functions

---

### 8.1

Before looking at these functions, a more elementary matter of nomenclature should be cleared up. It concerns the words 'power', 'index' and 'exponent'.

The words 'index' and 'exponent' are interchangeable. (There is a tendency to call a superscript an index if it is a number and an exponent if it is a letter, but it is quite unnecessary to preserve any such distinction.) But 'power' is *not* synonymous with 'index': $x^4$ and $x^{-3}$ are *powers* of $x$; the *indices* of those powers are 4 and $-3$.

A logarithm of a number to a given base is the *index of the power* to which that base must be raised for it to equal the given number. Thus, if the base is $b$ ($b$ is always positive),

$$\lg_b x = y \quad \Leftrightarrow \quad x = b^y:$$

$b^y$ is a power of $b$; $y$ is the index *of* that power.

Logarithm tables are tables of *indices*: antilogarithm tables are tables of *powers* (of 10). The terms 'index' and 'power' are in this sense inverse: they are certainly *not* interchangeable!

Note, in passing, that the cartesian graph of $y = \lg_b x$ only has the familiar form in Fig. 8.1 if $b > 1$. If $0 < b < 1$, it is located as shown in Fig. 8.2.

### 8.2

The natural logarithm arises in connexion with the problem of integrating $1/x$. The point that needs emphasizing here is that

$$\int \frac{1}{x}\,\mathrm{d}x = \begin{cases} \ln \quad x \ + \mathrm{A} & \text{if} \quad x > 0 \\ \ln(-x) + \mathrm{B} & \text{if} \quad x < 0. \end{cases} \tag{1}$$

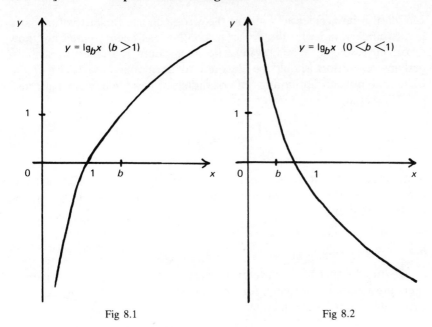

Fig 8.1                          Fig 8.2

Pupils should be encouraged never to omit the second half of this statement. This may help them to avoid the sort of nonsense they are apt to write when evaluating an integral such as

$$\int_1^2 \frac{dx}{x-4},$$

making ghastly howlers like

$$[\ln(x-4)]_1^2 = \ln(-2) - \ln(-3).$$

The fact that this 'method' may seem to be retrospectively justified when the correct answer, $\ln \frac{2}{3}$, appears does not make this terrible calculation respectable.

One not infrequently comes across a student who is competent at obtaining indefinite integrals of outlandish complexity, but who is stumped when he has to decide whether to take $\ln[f(x)]$ or $\ln[-f(x)]$ in order to evaluate, for example,

$$\int_a^b \frac{dx}{k^2 - x^2} \quad \text{according as} \quad \left\{ \begin{array}{c} a < b < -|k| \\ -|k| < a < b < \quad |k| \\ |k| < a < b, \end{array} \right.$$

and who, if he accidentally makes the wrong choice to start with, is at a loss to understand why the apparent conflict has been created and how it can be resolved. When 'systematic integration' is studied in schools, perhaps less effort should be devoted to developing the tricks of the trade and more to promoting understanding of basic behaviour patterns.

Telescoping the two statements (1) into

$$\int \frac{1}{x}\, dx = \ln|x| + C \qquad (2)$$

is admirable, but it is best not to suggest this simplification at once, otherwise its benefit is not appreciated and the modulus signs are too readily forgotten.

## 8.3

Textbooks that teach differential equations often leave large gaps in their presentation. In many, for example, you will see a solution of the first-order differential equation

$$\frac{dy}{dx} = ky \qquad (k \text{ constant}) \qquad (3)$$

that just says

$$\frac{1}{y}\frac{dy}{dx} = k \qquad (4)$$

$$\ln y = kx + A \qquad (5)$$

$$y = Be^{kx}. \qquad (6)$$

The final answer is, of course, correct and happens to be completely general.

But this traditional solution has tripped lightly across a whole minefield of assumptions, which the writer either did not think worth mentioning or perhaps did not even notice. The omission of the condition $y \neq 0$ in passing from (3) to (4) may perhaps be excused as being implied (although such laxity cannot be held to set a good example for readers to follow). But the step from (4) to (5) is only correct if $y > 0$: if $y < 0$, step (5) has to be replaced by $\ln(-y) = kx + A'$. Also, writing A as $\ln B$ in preparation for the final transition from (5) to (6) requires $B > 0$. Does that mean that B has to be positive in the solution (6)? Think about this.

To solve the differential equation (3) adequately, one must either consider separately the cases $y > 0$ and $y < 0$, or else replace line (5) by

$$\ln |y| = kx + A. \tag{7}$$

Putting $A = \ln B$ where $B > 0$ leads to

$$|y| = B\, e^{kx}. \tag{8}$$

Since $e^{kx} > 0$ for all $x$, $y$ can never change sign; in other words, $y$ must be of constant sign throughout the solution. Thus it is possible to simplify the solution (8) by *simultaneously* removing the modulus signs *and* allowing the constant $B$ to be negative as well as positive. Finally, by allowing $B$ to be zero so as to include the trivial solution $y = 0$, which has been excluded until now, the solution of (3) is obtained in all its generality as

$$y = B\, e^{kx},$$

where $B$ is unrestricted. An alert reader will often find similar examples of slipshod arguments in textbooks.

## 8.4

A related, but more subtle, conundrum can arise in the solution of other types of linear differential equation. Consider, for example, the equation

$$\frac{dy}{dx} + \frac{y}{1 - x^2} = \frac{x}{x + 1} \qquad (|x| \neq 1),$$

which you are invited to solve for yourself before proceeding.

According to whether you chose

$$\left(\frac{1 + x}{1 - x}\right)^{1/2} \quad \text{or} \quad \left(\frac{x + 1}{x - 1}\right)^{1/2}$$

for your integrating factor, you will have obtained either

$$y + 1 = x + A\left(\frac{1 - x}{1 + x}\right)^{1/2}$$

or

$$y + 1 = x + B\left(\frac{x-1}{x+1}\right)^{1/2}.$$

If you now recall the restrictions on $x$ that were lurking in the background when $\int dx/(1-x^2)$ was being evaluated [see Section 6.2], you will appreciate that neither of these solutions is complete by itself. Nevertheless, you would have to search long and hard to find a textbook that, in a case like this, included both the above possibilities, for $|x| < 1$ and for $|x| > 1$ respectively, in its answer.

A perhaps simpler equation to illustrate this same point is

$$\frac{dy}{dx} + \frac{xy}{1-x^2} = x \qquad (|x| \neq 1),$$

where the choice of $(1-x^2)^{-1/2}$ or $(x^2-1)^{-1/2}$ for the integrating factor leads to the partial solutions

$$y + 1 = x^2 + A(1-x^2)^{1/2} \qquad (|x| < 1)$$

and

$$y + 1 = x^2 + B(x^2-1)^{1/2} \qquad (|x| > 1),$$

which together make up the complete solution.

### 8.5

The exponential function of a real variable $(\mathbf{R} \to \mathbf{R} : \exp : x \mapsto e^x)$ does not call for any special comment here. But the problems involved in extending this theory to $\exp z (z \in \mathbf{C})$ will now be the subject of an extensive discussion. Even though this will range well beyond the usual school syllabus, it provides background information that is rather necessary, especially if you have any inclination to press $\exp z$ into service (as many teachers do) to solve real differential equations like

$$\frac{d^2y}{dx^2} - 6\frac{dy}{dx} + 13y = 0.$$

The exponential function of a complex variable is normally introduced by showing that the series $\Sigma z^n/n!$ absolutely converges for all complex $z$, so that its sum defines a *function*

$$\mathbf{C} \to \mathbf{C} : \exp,$$

where

$$\exp z = \sum_{0}^{\infty} \frac{z^n}{n!}.$$

Application of a standard theorem about the multiplication of absolutely convergent series then leads to the discovery that

$$\exp z_1 . \exp z_2 = \exp(z_1 + z_2) \quad \text{for all} \quad z_1, z_2 \in \mathbf{C}.$$

The proof of that standard theorem, however, is not particularly easy, certainly for school presentation, and itself depends on the important but rather subtle theorem about the rearrangement of terms in an absolutely convergent series.

This result, however, combined with the easily established observations that, if $t \in \mathbf{R}$,

$$\exp t = e^t \quad \text{and} \quad \exp(it) = \cos t + i \sin t, \quad \text{or} \quad \text{cis} \, t \quad \text{for short,}$$

is needed to obtain the corollary that

$$\exp(u + iv) = e^u \text{cis} \, v \quad \text{for} \quad u, v \in \mathbf{R}.$$

Since $e^u > 0$ for all $u \in \mathbf{R}$, $\exp(u + iv)$ is never zero, but all non-zero complex numbers arise as images under exp and so the range of the function exp is $\mathbf{C} \setminus \{0\}$.

## 8.6

Starting from the relation $z = \exp w$, it is reasonable, when $z \, (\neq 0)$ is given, to regard any possible value of $w$ as a logarithm of $z$. Now, if $w = u + iv, z = e^u \text{cis} \, v$ and hence $e^u = |z|$ and $v = \text{Arg} \, z$,[†] so that

$$w = \ln |z| + i \, \text{Arg} \, z.$$

Thus there is an unbounded number of values of $w$ and so, in conformity with the convention about $\text{Tan}^{-1}$, etc., discussed in the previous chapter, the relation $z = \exp w$ will be expressed in the equivalent form[‡]

$$w = \text{Ln} \, z,$$

[†]See Section 7.3 for the notation 'Arg'.
[‡]It would be quite consistent to write $\text{Exp}^{-1}$ instead of Ln (giving an even closer analogy with $\text{Tan}^{-1}$), but the word 'logarithm' is too well established to warrant this.

with an upper case L, so that

$$\text{Ln } z = \ln |z| + i \text{ Arg } z \qquad (z \neq 0).$$

A logarithm *function*

$$\mathbf{C} \setminus \{0\} \to \mathbf{C} : \ln$$

(written with lower case l) can then be defined using the function arg [see Section 7.3], so that

$$\ln z = \ln |z| + i \arg z \qquad (z \neq 0).$$

With the same interpretation as that used in the previous chapter in connexion with Arg, you can, if you wish, make statements like

$$\text{Ln}(z_1 z_2) = \text{Ln } z_1 + \text{Ln } z_2$$

$$\text{Ln } \frac{1}{z} = -\text{Ln } z$$

and so on, but, if you do, you must stress that, in contrast,

$$\ln z_1 + \ln z_2 \quad \text{is not necessarily equal to} \quad \ln(z_1 z_2).$$

[Before proceeding further, it must regretfully be pointed out that the authors of several books on complex analysis (including the Open University units on this subject) have misguidedly chosen to use Ln and ln (or Log and log) in senses precisely *opposite* to the meanings given above. Apart from the fact that such usage runs counter to the general principles outlined in Section 7.1 that function symbols have small initial letters while capital letters refer to relations, the main objection to this absurd choice is that, when $z$ is $x$-axal with $x$ positive, it is then $\text{Ln}(x + 0i)$ that is the same as $\ln x$, whereas $\ln(x + 0i)$ also occurs but means something different! Indeed, to these writers, the sum of the series

$$x - \tfrac{1}{2}x^2 + \tfrac{1}{3}x^3 - \quad \ldots \quad (-1 < x \leqslant 1)$$

is $\ln(1 + x)$ if $x \in \mathbf{R}$ and $\text{Ln}(1 + x)$ if $x + 0i \in \mathbf{C}$. How stupid can one be in selecting one's notation? It has been noted in Section 3.9 (when considering $x^{m/n}$ and $(x + 0i)^{m/n}$ and contrasting these with $|x|$ and $|x + 0i|$), that, because mathematicians have adopted a bad notation in the first place, it is not always practicable to ensure that $f(x + 0i)$ is exactly the same as $f(x)$, but it is obviously an objective worth pursuing unless the difficulties are insuperable. But to have $\ln(x + 0i)$ different from $\ln x$ and then $\text{Ln}(x + 0i)$ the *same* as $\ln x$ is doubly irresponsible. The *natural* convention of using ln for the function and Ln in the relation ensures that $\ln x$ $(x > 0)$ has exactly the same meaning whether $x$ is a real number or an $x$-axal complex number. Books that use the illogical opposite convention should be resolutely avoided.]

### 8.7

When the domain is $\mathbf{C} \setminus \{0\}$, that is, the complex plane with just the origin removed, the function ln is *not* continuous throughout the domain, because of the discontinuity of the function arg along the negative $x$-axis. As $z = x + iy$ tends to $-1$ through values for which $y > 0$, $\ln z \to i\pi$; but as $z$ tends to $-1$ through values for which $y < 0$, $\ln z \to -i\pi$. This discontinuity is an inherent feature of the logarithm. A redefinition of the bounds of arg $z$ could be used to move the position of these points of discontinuity, but they can only be shifted, never removed. These observations are of particular importance for the discussion on powers in Chapter 9.

To make the function ln continuous everywhere, some restriction must be imposed on its domain; some simple curve must be drawn joining the points 0 and $\infty$† and the points of this curve cut from the plane. The cut, in practice, is chosen to be along the non-positive real axis and the plane from which the origin and negative real axis have been removed is called the *cut plane*. When its domain is this cut plane, the function ln is continuous (indeed, analytic) throughout.

There are various reasons for this choice of cut. One certainly wants to avoid the positive real axis, because one particularly wishes the function to be continuous there, where $\ln(x + 0i)$ coincides with $\ln x$, the value taken by the ordinary real function. [It is this sort of consideration to which the writers censured in the paragraph in brackets above are obviously insensitive!] This observation, incidentally, also explains why the definition of arg $z$ is taken to be

$$-\pi < \arg z \leqslant \pi$$

and *not* $0 \leqslant \arg z < 2\pi$, as a pupil starting complex algebra might expect.

---

†See Chapter 4 for a discussion of why one *can* legitimately talk about the point $\infty$ in the complex plane, when one can *not* talk about a real number $\infty$.

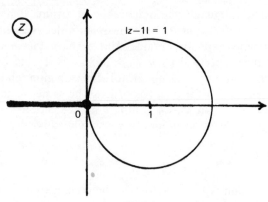

Fig 8.3

More generally, however, the Taylor series

$$\zeta - \frac{\zeta^2}{2} + \frac{\zeta^3}{3} - \ldots + (-1)^{n-1}\frac{\zeta^n}{n} + \ldots$$

converges for all $\zeta$ satisfying $|\zeta| \leq 1$, *except* for $\zeta = -1$, and one wishes to write the sum as $\ln(1 + \zeta)$ for *all* these values of $\zeta$. One must, therefore, avoid taking a cut for $\ln z$ that intersects the region $|z - 1| \leq 1$ anywhere except at $z = 0$ (Fig. 8.3). This offers the straight

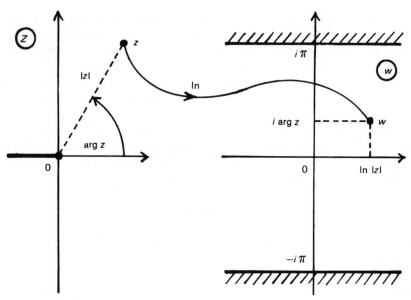

Fig 8.4

cut along the non-positive $x$-axis as the simplest and most 'natural' choice. It has the effect that, when $w = \ln z$, the image of the cut $z$-plane in the $w$-plane (that is, the range of ln for that domain) is a strip of width $2\pi i$ symmetrically placed about the real axis in the $w$-plane, as shown in Fig. 8.4.

## 8.8

There is another problem associated with the exponential notation. Some authors write $\exp z$ as $e^z$ and many that do not do this for general $z$ do it at least when $z$ is $y$-axal, writing $\exp it$ ($t \in \mathbf{R}$) as $e^{it}$ and saying that $\operatorname{cis} t = e^{it}$. This usage has been sanctioned by many illustrious mathematicians, including Euler and Gauss, no less. Most teachers will not hesitate to follow such eminent authorities, yet it is worth pointing out that, from an educational point of view, there may be advantages of consistency in not doing so.

As explained above, exp is incontrovertibly defined as a *function*; it has the property that, when $x$ is *real*, $\exp x$ is exactly the same as $e^x$. The problem of giving $e^z$ a meaning for general $z$ is a special case of that of defining $\zeta^z$ when $\zeta$ and $z$ are both complex and $\zeta \neq 0$. This topic is covered in the next chapter, where it is shown that the definition of $\zeta^z$ (and, in particular, of $e^z$) that seems most natural allots it an unbounded number of values (in general). One of these values of $e^z$ turns out to be $\exp z$, but to use $e^z$ interchangeably with $\exp z$ prevents a possible distinction from being sustained. There are, as discussed in Chapter 3 on square roots and again in Chapter 9, difficulties with the power notation whatever convention is adopted, since mathematicians perversely insist on making do with the same (inefficient) symbolism to represent the outcome of two related but distinguishable processes.

Whatever your reaction to the general arguments in the next chapter, there is a strong case for using only $\exp z$ (never $e^z$) for the image of $z$ under the exponential mapping and, in fact, replacing $\exp z$ by $e^x$ *only* when $z$ ($= x + 0i$) is $x$-axal. This leaves the symbol $e^z$ free, if required, for a different interpretation. This stricter point of view will be adopted here. Thus, for example, in the solution below, $y = \exp mx$ has been written, and *not* $e^{mx}$, because complex values of $m$ are going to be accepted, even though $x$ is real.

## 8.9

As said earlier, it is quite common for teachers to use complex exponentials for solving (real) differential equations, even if this is the only application they make of this theory. It is very doubtful whether it is really worth while introducing one's pupils to exp $z$, if this is to be the sole reason for studying the subject. For, as you will see when you inspect the solutions offered, they normally contain unplugged loopholes anyway.

School textbooks that recommend this technique usually solve an equation like

$$\frac{d^2y}{dx^2} - 6\frac{dy}{dx} + 13y = 0 \tag{9}$$

by saying that $y = \exp mx$ is a solution if and only if

$$m^2 - 6m + 13 = 0, \tag{10}$$

that is,

$$(m - 3)^2 + 4 = 0, \tag{11}$$

or

$$m = 3 \pm 2i. \tag{12}$$

Hence the solution is

$$y = e^{3x}[A'\exp(2ix) + B'\exp(-2ix)], \tag{13}$$

or

$$y = e^{3x}(A\cos 2x + B\sin 2x). \tag{14}$$

The step from (13) to (14) is no problem: the association of exp $it$ for $t \in \mathbf{R}$ with $\cos t + i\sin t$ is easy to justify. But that from (12) to (13) presents greater difficulty. Replacing $\exp(u + iv)$ by $e^u\exp iv$ for $u$, $v \in \mathbf{R}$ depends on the identity $\exp(z_1 + z_2) = \exp z_1 . \exp z_2$ and, as already pointed out, this is not particularly easy to establish and so is usually taken as an unproved assertion.

But there is a further gap: and this one usually passes completely unnoticed. Step (10) depends on the result

$$\frac{d}{dx}(\exp mx) = m\exp mx, \tag{15}$$

where $m$ may be real or complex.

All that has usually been *proved* in the book is that

$$\frac{d}{dx}(e^{mx}) = me^{mx} \quad \text{when} \quad m \text{ is } real.$$

Although there may be some hope that the corresponding result (15) will hold for exp $mx$ when $m$ is complex, there is certainly no guarantee, without proper investigation, that this will be the case.

To justify (15), you must at all costs avoid attempting to derive [differentiate: see Chapter 10] a series term-by-term. As observed in Chapter 13, even for a real series this is a hazardous undertaking, not to be embarked on lightly: for a complex series, the difficulties are compounded.

The starting point for an elementary proof can be the result that, if $\mathbf{R} \rightarrow \mathbf{C} : w$ is given and

$$w(x) = u(x) + iv(x),$$

where $\mathbf{R} \rightarrow \mathbf{R} : u$ and $\mathbf{R} \rightarrow \mathbf{R} : v$ are real-valued derivable functions of a *real* variable, then

$$w'(x) = u'(x) + iv'(x).$$

This is proved in the obvious way, just by examining $w(x + \delta x) - w(x)$; fortunately, the more tricky problem of defining $w'(z)$ when $\mathbf{C} \rightarrow \mathbf{C} : w$ is a function of a *complex* variable can be by-passed.

Taking

$$w(x) = \exp mx, \quad \text{where} \quad m = a + ib \qquad (a, b \in \mathbf{R}),$$
$$w(x) = e^{ax}\cos bx + i\,e^{ax}\sin bx.$$

So

$$\begin{aligned} w'(x) &= e^{ax}(a\cos bx - b\sin bx) + i\,e^{ax}(a\sin bx + b\cos bx) \\ &= (a + ib)e^{ax}(\cos bx + i\sin bx) \\ &= m\exp mx, \end{aligned}$$

establishing the result (15) required and so plugging the gap in the above 'solution'.

But the differential equation (9) can easily be solved by reducing it to one of simple harmonic type, whose solution will be a standard piece of work for these pupils anyway. The substitution

$$y = e^{3x}Y$$

transforms (9) to

$$\frac{d^2Y}{dx^2} + 4Y = 0,$$

—compare (11)—whose solution is

$$Y = A \cos 2x + B \sin 2x,$$

giving (14) as before.

A person who argues that he teaches $\exp z$ because of its application to solving differential equations has a very poor case and, if this is his only reason for introducing the concept, he is building a steam-hammer to crack a peanut. The time can be much better spent on other topics, particularly since this technique of solution is often, rather surprisingly, advocated for the weak student.

## 8.10

A better case can be advanced for defining $\exp z$ with the able boys and girls. The objective there would be the different one of expanding the pupils' horizons. The properties of the complex exponential function reveal the reasons for the analogies between the circular and hyperbolic functions, explain the similarities between their addition formulae, Osborn's rule and so on, whereas the fact that its period, $2\pi i$, is $y$-axal largely determines the differences between the two groups. Any application to real differential equations would be incidental to this purpose: just another rather interesting use of complex numbers in solving a real variable problem.

Such discussion has the bonus of showing them that, apart from polynomials and rational functions, every one of the 26 functions

> exp, sin, cos, tan, cot, sec, csc, sh, ch, th, coth, sech, csch,
> ln,  $\sin^{-1}$,       . . .       , $\text{sh}^{-1}$,       . . .       ,

they meet at school, as well as general power functions $(x \mapsto b^x)$, general logarithms $(x \mapsto \lg_b x)$ and relations involving $\text{Sin}^{-1}$, $\text{Ch}^{-1}$, etc. are all really aspects of a single entity: the complex exponential function. According to their temperament, students may find this discovery either stimulating or frustrating!

# Chapter 9

# The notation for powers

---

## 9.1

The successive definitions of the power $b^a$ when the index $a$ and base $b$ belong to various number systems punctuate the school curriculum as pupils' experience widens. This recurring topic is a perfect paradigm of one important feature of mathematics—the constant striving for greater generality—and can usefully be exploited to make students aware of this aspect of the subject.

The process starts quite soon after the beginning of algebra, when the index is a positive integer, $m$, and $b^m$ is defined as the product of $m$ factors each equal to $b$. [Incidentally, the result of 'multiplying $b$ by itself $m$ times', which one not infrequently hears, is surely $b^{m+1}$, is it not?]

Before long the very important question arises of trying to extend the definition of $b^a$ to rational $a$. For many pupils, this may be a turning point in their mathematics education and it needs sensitive handling by the teacher. For the first time, the children are being faced with a really abstract definition, not suggested by experience or undertaken merely for convenience. It is not just a piece of artificial symbolism that is being introduced (like the shorthand $b^4$ for $bbbb$ or the arbitrary use of parentheses to distinguish $x(y + z)$ from $xy + z$), which is what most of their previous algebra has been: it is a *genuine mathematical* definition. Intuition does not suggest a meaning for $b^{-5}$ or $b^{2/3}$, nor even whether it is likely that a sensible and useful definition can be made. The boys and girls are being put into the position of authentic mathematicians and are being initiated into what the subject is all about. We do not want them to falter at this hurdle or we may lose them for good.

They are first invited to examine the properties of symbols they are already familiar with (in this case, $b^m$ with $m \in \mathbf{P}$, the positive integers) and to see how they behave (by compiling the 'index laws'). They then have to search this schedule for clues as to whether it would be sensible to try and *give* meanings to some new quantities ($b^{-m}$, $b^0$, $b^{m/n}$) and, if so, what those meanings ought to be. Now comes the crucial step: the recognition that the meanings to be given to these new symbols are matters of *definition, not* of proof. Finally, there is the examination of the properties of the newly introduced symbols to check that the previous scheme ($b^m$ with $m \in \mathbf{P}$) fits harmoniously into the new ($b^a$ with $a \in \mathbf{Q}$) without inconsistency.

This is one of the classic ways in which all branches of mathematics develop. A concept has been defined with a certain range of application but the definition is meaningless outside that territory. One looks to see if one can find some property associated with the concept that (1) will make sense outside the ambit of the definition, (2) can be used to extend the scope of the idea beyond its original compass and (3) will do this in such a way that the new concept will agree with the old whenever both interpretations are appropriate.

[As a more advanced example, consider the way in which the factorial function is extended. The original definition of $n!$ ($n \in \mathbf{P}$) could hardly be more dependent on $n$ being a positive integer. Yet it is extended first to $\mathbf{N}$, the natural numbers, by including 0 in the domain ($0! = 1$); then to all real numbers greater than $-1$ via the gamma function ($\Gamma(n + 1) = n!$), by using a certain integral, which happens to converge iff $x > 0$, to define $\Gamma(x)$. The functional relation $\Gamma(x + 1) = x \, \Gamma(x)$, which is *proved* for $x > 0$, is then used to provide the *definition* of $\Gamma(x)$ when $x < 0$. Eventually, by making a further leap, $\Gamma(z)$ is defined for all $z$ belonging to a suitable complex domain.]

So now $b^a$ is defined for all $a \in \mathbf{Q}$ and for $b > 0$. (Problems with $b < 0$ are avoided at this level by restricting $b$.) When $a = m/n$ and $n$ is even, $b^{m/n}$ is used at this stage to denote the *positive* $n$th root of $b^m$ (so that, in particular, $b^{1/2}$ is exactly the same as $\sqrt{b}$). [See the discussion on this point in Section 3.9.]


**9.2**

After the pupils have met irrational numbers, they may possibly inquire whether this procedure can be extended to suggest a definition for, say, $b^{\sqrt{2}}$, so it is as well to be prepared for this. In fact, of course, it cannot, but a certain amount of experimentation will be needed before they convince themselves that neither intuition *nor* the index laws offer any guidance here. It is, however, possible to approach the problem in a different way, provided you first establish the results that, when $a_1$, $a_2 \in \mathbf{Q}$,

$$\text{if} \qquad b > 1, \qquad a_1 < a_2 \quad \Rightarrow \quad b^{a_1} < b^{a_2};$$

$$\text{if} \quad 0 < b < 1, \quad a_1 < a_2 \quad \Rightarrow \quad b^{a_1} > b^{a_2}.$$

Two sequences of powers of $b$ with rational indices are constructed. Each sequence of indices tends steadily to $\sqrt{2}$ but from opposite sides: for example,

$$b^1, b^{1.4}, b^{1.41}, b^{1.414}, b^{1.4142}, \ldots$$

and

$$b^2, b^{1.5}, b^{1.42}, b^{1.415}, b^{1.4143}, \ldots$$

Although it is difficult at this level to justify rigorously that these sequences have a common limit (which limit it is reasonable to use as the definition of $b^{\sqrt{2}}$), the procedure is intuitively very persuasive and should lead to all sorts of stimulating discussion with interested pupils about the problems it raises. It may even be a good idea with your brightest classes to contrive to initiate such an investigation before elementary logarithms are introduced.

The more straightforward approach to $b^{\sqrt{2}}$ comes later, after the exponential and logarithmic functions have appeared on the scene. However they are introduced, $x \mapsto e^x$ emerges as a function with domain $\mathbf{R}$ and $x \mapsto \ln x$ as a function with domain $\{x \in \mathbf{R} : x > 0\}$, so that $b^a$ can be defined for all real $a$ and for all real, *positive* $b$ by

$$b^a = e^{a \ln b}.$$

The next step is to check that the properties of this symbol are consonant with all the earlier definitions and 'laws'.

## 9.3

At about the same time as this introduction to $e$ is taking place, the pupils will probably be broadening their knowledge in another direction by becoming acquainted with complex numbers. There they will discover how complicated the behaviour of the familiar functions sin and cos is when compared with the beautiful simplicity of the properties enjoyed by cis $[= \cos + i \sin]$. When it is set alongside the clumsiness of the addition formulae, they will enjoy savouring the austere elegance of de Moivre's theorem, whether this is given the arresting enunciation that

for all $a \in \mathbf{Q}$, cis $a\phi$ is one value of $(\text{cis } \phi)^a$, or the more general statement that

(cis $\phi$)$^{m/n}$ has exactly $n$ distinct values given, for example, by

$$(\text{cis } \phi)^{\frac{m}{n}} = \text{cis}\left(\frac{m}{n}\phi + \frac{2k\pi}{n}\right) \quad \text{for} \quad k = 0, 1, \ldots, n - 1.$$

Here, as discussed in Chapter 3, $\zeta^{m/n}$ ($\zeta \neq 0$, $m$ and $n$ coprime) is being used to represent any number $w$ satisfying $w^n = \zeta^m$ and so $\zeta^{m/n}$ is any element belonging to a *set* of $n$ complex numbers, in which no individual is preferred above the rest, and whose geometric configuration is particularly striking: the vertices of a *regular* polygon. This definition means, in particular, that when $\zeta = b + 0i$ where $b \in \mathbf{R}$, $(b + 0i)^{1/2}$ denotes a pair of numbers, in distinction to the meaning quoted above for $b^{1/2}$ when $b > 0$.

As noted in Section 8.6, ln is defined as a *function* of a complex variable with domain $\mathbf{C} \setminus \{0\}$, and so it is natural (by analogy with $b^a = e^{a \ln b}$ when $b > 0$) to investigate

$$w = \exp\left(\frac{m}{n} \ln \zeta\right) \quad (\zeta \neq 0).$$

This certainly defines a particular complex number for all $\zeta \neq 0$ and hence $w$ is a function of $\zeta$ for that domain. Furthermore, $w$ does have the property that

$$w^n = \exp\left(m \ln \zeta\right) = \zeta^m,$$

so that $w$ is indeed one of the values of $\zeta^{m/n}$, and this appears to contradict our earlier claim by selecting a distinguished element of the set.

The trouble with this idea, however, is that $w$ is *not* a *continuous* function of $\zeta$ in the domain $\mathbf{C} \setminus \{0\}$. When, for instance, $\zeta \to -1$ through values of $\zeta = \xi + i\eta$ for which $\eta > 0$,

$$\exp\left(\frac{m}{n} \ln \zeta\right) \to \text{cis}\left(\frac{m}{n}\pi\right),$$

whereas, when $\zeta \to -1$ through values of $\zeta$ for which $\eta < 0$,

$$\exp\left(\frac{m}{n} \ln \zeta\right) \to \text{cis}\left(-\frac{m}{n}\pi\right);$$

for example, if $m/n = 1/2$, $w \to i$ and $w \to -i$ respectively.

This was what was meant when it was said that a distinguished member could not be picked out consistently from the set. What, more formally and correctly, was meant was that the number selected could not be a continuous function of $\zeta$ throughout $\mathbf{C} \setminus \{0\}$; or, even more precisely, that, when $\zeta$ describes a continous closed curve surrounding the origin, it is not possible (except when $n = 1$) to arrange for the selected value also to describe a continuous *closed* curve—and this would be essential if the selection were to have any value.

This property is an intrinsic feature of the logarithm function, as explained in Section 8.7. The function ln can only be made continuous by taking a more restricted domain, namely, the cut plane (from which the non-positive real axis has been excised). But that would not have helped above, since the objective there was to make a choice for *all* $\zeta \neq 0$.

### 9.4

When it comes to giving a meaning to $\zeta^z$ for general complex numbers $\zeta$ and $z$, it is natural, if the spirit of these earlier definitions is to be followed, to give $\zeta^z$ the widest possible interpretation and agree that, when $\zeta \neq 0$,

$$\zeta^z \text{ will denote } any \text{ value of } \exp(z \operatorname{Ln} \zeta). \tag{1}$$

Then, if $z = x + iy$ and $\zeta = \rho$ cis $\phi$ $(\rho > 0)$, $\operatorname{Ln} \zeta = \ln \rho + i\phi$ [see Chapter 8] and

$$z \operatorname{Ln} \zeta = (x + iy)(\ln \rho + i\phi) = (x \ln \rho - y\, \phi) + i(y \ln \rho + x\, \phi).$$

Hence

$$\zeta^z = e^{x \ln \rho - y\, \phi} \operatorname{cis}(y \ln \rho + x\, \phi) = \rho^x\, e^{-y\phi} \operatorname{cis}(y \ln \rho)\operatorname{cis}(x\, \phi). \tag{2}$$

Here $\phi$ is *any* value of Arg $\zeta$, although the same value must, of course, be taken for $\phi$ at each occurrence. In general, the number of values of $\zeta^z$ will be unbounded.

By taking $\phi = \arg \zeta$ (which is equivalent to replacing $\operatorname{Ln} \zeta$ in (1) by $\ln \zeta$), it would be possible to pick out one particular value of $\zeta^z$, but the discontinuity of arg is just as insuperable an obstacle to doing this consistently in the general case as it was in the special case $z = m/n$ just discussed.

Incidentally, it would be prudent at this point to check that $\zeta^{m/n}$ has not acquired any extra values with this definition (1), beyond those that were assigned to it earlier. According to the new definition, for all $\zeta \neq 0$,

$$\zeta^{m/n} = \exp\left(\frac{m}{n} \operatorname{Ln} \zeta\right) = \exp\left[\frac{m}{n} (\ln|\zeta| + i \operatorname{Arg} \zeta)\right] = |\zeta|^{m/n} \operatorname{cis}\left(\frac{m}{n} \operatorname{Arg} \zeta\right).$$

It is easily seen that this formula gives exactly the same set of $n$ values as that prescribed by de Moivre's theorem. [For, if $\operatorname{Arg} \zeta = \phi + 2h\pi$,

$$\frac{m}{n} \operatorname{Arg} \zeta = \frac{m}{n} \phi + \frac{2hm\pi}{n}.$$

If $hm \equiv k(n)$, (that is, for some integers $k, l$, $hm = k + ln$), then

$$\frac{2hm\pi}{n} = \frac{2k\pi}{n} + 2l\pi$$

and so

$$\operatorname{cis}\left(\frac{m}{n} \phi + \frac{2hm\pi}{n}\right) = \operatorname{cis}\left(\frac{m}{n} \phi + \frac{2k\pi}{n}\right).$$

But, if $h_1 m \equiv k_1(n)$ and $h_2 m \equiv k_2(n)$, then, since $m$ and $n$ are coprime,

$$h_1 \equiv h_2(n) \qquad \Leftrightarrow \qquad k_1 \equiv k_2(n).$$

Thus neither formula gives any additional values of $\zeta^{m/n}$ not contained in the other.]

So far, any definition of $0^z$ has been avoided, because the procedure involving logarithms would not apply. In fact, if $z = x + iy$, this symbol can be defined if *and only if* $x > 0$, in which case, $0^z$ can be given the value 0. To see that this is reasonable, refer to equation (2) above and consider the circumstances under which $\zeta^z$ will tend to a limit as $\rho \searrow 0$.

The meaning the definition (1) gives to $e^z$ will now be investigated. In fact,

$$e^z = \exp(z \operatorname{Ln} e) = \exp[z(1 + 2k\pi i)] = \exp z \,.\, \exp(2k\pi iz),$$

where $k$ is an integer. Hence $\exp z$ is just *one* of the values of $e^z$ and so by *not* writing $\exp z$ as $e^z$ it is possible to maintain a distinction between these two. If $z = x + iy$,

$$e^{x+iy} = e^{x - 2k\pi y} \operatorname{cis}(y + 2k\pi x).$$

In particular,

$$e^{it} = e^{-2k\pi t} \operatorname{cis} t,$$

so that, when the suggested distinction is made, cis $t$ or exp $it$ is just *one* value of $e^{it}$. This is why many teachers prefer not to write $\cos t + i \sin t$ as $e^{it}$, but only as exp $it$.

Similar remarks apply to $b^z$ for *all* $b$ that are *real* and *positive*.

$$b^z = \exp(z \operatorname{Ln} b) = \exp[z(\ln b + 2k\pi i)] = \exp(z \ln b)\exp(2k\pi i z);$$

$$b^{x+iy} = e^{x \ln b} e^{-2k\pi y} \operatorname{cis}(y \ln b + 2k\pi x).$$

In all these cases, a particular value of $b^z$, namely $\exp(z \ln b)$, *is* distinguished from all the others and this happens in a very dramatic and important way.

## 9.5

The reason why this selection process *is* possible here, in contrast to what has already been observed in the general case, will be clarified presently, but first there are two other observations worth making.

When $y = 0$,

$$b^{x+0i} = e^{x \ln b} \operatorname{cis} 2k\pi x,$$

so that, if $b$ is real and positive, $e^{x \ln b}$, which by definition is *the* value of $b^x$ in real algebra, becomes just *one* of the values of $b^{x+0i}$ (albeit the distinguished one) in complex algebra. So here the symbol $b^x$ may be ambiguous and this is another place where $f(x)$, interpreted as $f(x + 0i)$, has a meaning different from $f(x)$, $x$ real. [Altering the definition in this case by using $b^z$ to mean $\exp(z \ln b)$, the special value, rather than $\exp(z \operatorname{Ln} b)$, the general value, would *not* resolve the conflict, of course, because that would merely mean that $b^z$ ($b$ real and positive) was being used differently from $\zeta^z$ ($\zeta$ complex and non-zero) where this selection procedure can *not* operate consistently.] What a pity the power notation was chosen as it was; it makes mathematics so untidy!

The other important piece of evidence concerns the binomial series. The real expansion, which many pupils will meet at school, has a very straightforward generalization in complex algebra. Both the real and complex results are stated below in their full generality, although the first enunciation is. of course, included in the second.

Let $x$ be real and let $a$ be any real number other than a non-negative integer (for which case, the binomial series terminates). If

$$(1) \quad |x| < 1 \qquad\qquad \text{(this is the important case)}$$

$$\text{or} \quad (2) \quad x = 1 \quad \text{and} \quad a > -1$$

$$\text{or} \quad (3) \quad x = -1 \quad \text{and} \quad a > 0,$$

then the series

$$1 + ax + \frac{a(a-1)}{2!} x^2 + \cdots + \frac{a(a-1)\ldots(a-n+1)}{n!} x^n + \ldots$$

converges and the sum is the real *positive* value of $(1+x)^a$, the qualification emphasized being relevant if $a$ is rational and has an even denominator. This statement is correct for irrational $a$ provided $(1+x)^a$ is defined as $e^{a \ln(1+x)}$ when $x > -1$, and as 0 in case (3) when $x = -1$ and $a > 0$. [This definition, by the way, is in complete harmony with the previous interpretation, as it automatically picks out the *positive* value of $(1+x)^a$ when $a$ is a rational number with an even denominator.]

Now let $z$ and $a$ be complex and let $\mathscr{X}(a)$ denote the x-axal part of $a$. If

$$(1) \qquad\qquad |z| < 1$$

$$\text{or} \quad (2) \quad |z| = 1 \quad \text{but} \quad z \neq -1 \quad \text{and} \quad \mathscr{X}(a) > -1$$

$$\text{or} \quad (3) \qquad\qquad z = -1 \qquad\qquad \text{and} \quad \mathscr{X}(a) > 0,$$

then the series

$$1 + az + \frac{a(a-1)}{2!} z^2 + \ldots + \frac{a(a-1)\ldots(a-n+1)}{n!} z^n + \ldots$$

converges and the sum is

$$\exp[a \ln(1+z)] \text{ in cases (1) and (2) and 0 in case (3).}$$

It is naturally very tempting to write this sum as $(1+z)^a$, and, indeed, this is normally done. But observe that it does cause notational confusion. If, for example, $a$ is rational $(a = m/n$, say), then $\exp[a \ln(1+z)]$ is *one* of the $n$ values of $(1+z)^a$ as this symbol may be used elswhere; so that, when in this context the sum is written as $(1+z)^a$, that expression is being given a meaning different from the other one.

Miraculously, however, the difficulty discussed earlier about making a systematic selection, stemming from the discontinuity of the logarithm function, does not arise in this case. It *is* possible to pick out consistently one of the $n$ values (the one that is equal to the sum of the binomial series) and to keep it distinguished from all the others no matter how $z$ varies.

The reason for this is to be found in the restriction on $z$. The point $b = 1 + z$ lies somewhere inside the circle shown in Fig. 9.1, or in certain circumstances on the circle, but never outside. In cases (1) and (2), the intersection of this domain for $b = 1 + z$ and the cut along the non-positive real axis in the $b$-plane is empty, so the discontinuities are avoided and the selection of the distinguished value *is* possible.

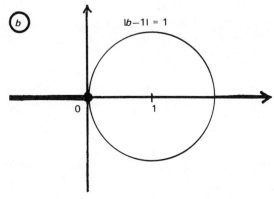

Fig 9.1

## 9.6

It will be instructive to illustrate this by looking at the associated geometry in the complex plane in a little more detail when $a$ is rational. Only the two simplest cases, $a = \pm\frac{1}{2}$, will be mentioned here, but you will find it an interesting exercise to examine other values of $a$ for yourself.

Let

$$\exp(a \ln b) = r \operatorname{cis} \theta$$

be the distinguished value of $b^a$, where $b = 1 + z$ and $|z| \leqslant 1$, $z \neq -1$.
    If $a = \frac{1}{2}$,

$$b = r^2 \operatorname{cis} 2\theta,$$

where $r \neq 0$ and, since $2\theta$ is to be arg $b$, $-\frac{1}{2}\pi < 2\theta < \frac{1}{2}\pi$. Thus

$$1 \geqslant |b - 1|^2$$
$$= |(r^2 \cos 2\theta - 1) + i(r^2 \sin 2\theta)|^2$$
$$= r^4 - 2r^2 \cos 2\theta + 1.$$

Hence

$$r^2 \leqslant 2 \cos 2\theta \qquad (-\tfrac{1}{4}\pi < \theta < \tfrac{1}{4}\pi).$$

Thus the point representing the distinguished value of $b^{1/2}$ always lies inside (or on) *one specific lobe* of the leminiscate $r^2 = 2 \cos 2\theta$, shown in Fig. 9.2.

It is this breaking up of the complete locus of $w$ (where $w^2 = b$) into two naturally separate regions that makes possible the consistent

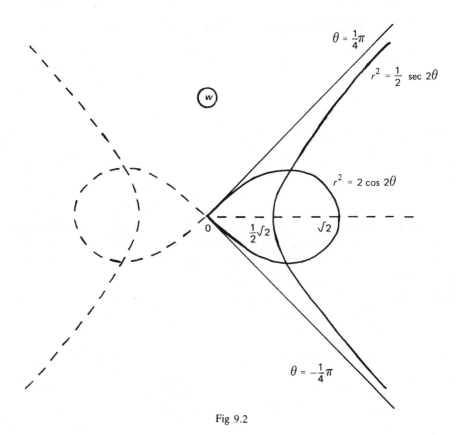

Fig 9.2

discrimination between the two values of $w$, with the distinguished value, $\exp(\frac{1}{2}\ln b)$, of $b^{1/2}$ lying exclusively within the right lobe of the leminiscate and the other value $[-\exp(\frac{1}{2}\ln b)$, corresponding to $\frac{3}{2}\pi < 2\theta < \frac{5}{2}\pi]$, confined to the left lobe.

This natural cleavage of the locus of $w$ is an automatic consequence of $b$ being restricted to the cut plane and can be contrasted with what happens in a more general case. If, for example, $|b| \leqslant 9$, then the resulting locus of the *pair* of points $w$ satisfying $w^2 = b$ is the disk $|w| \leqslant 3$. But this disk has *no* natural division into two subregions as the area bounded by the leminiscate has: nor can the possibility of one of these points $w$ being permanently distinguished from the other now arise.

Returning to the original domain for $b$ and taking $a = -\frac{1}{2}$, the distinguished value of $b^{-1/2}$ satisfies

$$r^2 \geqslant \tfrac{1}{2}\sec 2\theta \qquad (-\tfrac{1}{4}\pi < \theta < \tfrac{1}{4}\pi).$$

The point representing this value always lies beyond (or on) *one specific branch* of the rectangular hyperbola $r^2 = \frac{1}{2}\sec 2\theta$, also shown in Fig. 9.2. The natural separation of the locus of $w$ (where $w^2 = b^{-1}$) into two disjoint regions is again apparent. This geometric observation has its algebraic counterpart in the possibility of picking out a distinguished value of $b^{-1/2}$, by choosing the associated point $w$ exclusively from one region.

## 9.7

Thus, when the domain D of $b$ is sufficiently restricted (in fact, when it is a subregion of the *cut* plane), it *is* possible to select a particular value of $b^a$ in a consistent way, so that, for any $a \in C$, the chosen value is a continuous function of $b$ for all $b \in D$. In such cases, you will find that mathematicians often use $b^a$ for that chosen value, $\exp(a\ln b)$, even though it clashes with their other use of $b^a$ for *any* element of the complete *set* of values, $\exp(a \operatorname{Ln} b)$.

The truth should now be clear. There are two separate (if related) concepts and, with a false sense of economy, the same notation is made to serve for the representation of both. Obviously, mathematicians ought to have evolved different notations for these different purposes, but, as so often, they muddle along, to the bewilderment of every generation of students.

For purposes of discrimination *here*, two new symbols $b\uparrow^a$ and $b\Uparrow^a$ will be introduced to denote these two meanings of $b^a$. The symbol $b\uparrow^a$ will be used exclusively for $\exp(a\ln b)$ and will be *restricted* to contexts in which such an idea is viable, namely, when $b$ *is strictly excluded from*

**Table 9.1**

*Use of $b^a$ to mean $b \uparrow^a$.*

(1) $b \uparrow^a$ can be defined for all $a \in \mathbf{C}$ *provided $b$ is confined to the cut plane*; that is,

$$b \in \mathbf{C} \quad \Rightarrow \quad b \neq x + 0i \quad \text{with} \quad x \in \mathbf{R}, x \leq 0,$$

(so that, in particular,

$$b \in \mathbf{R} \quad \Rightarrow \quad b > 0).$$

(2) In that case,

$$\underline{b \uparrow^a = \exp(a \ln b).}$$

(3) The notation can, however, be extended to include $b = 0$ provided that the $x$-axal part of $a$ is then *positive*; otherwise, $0 \uparrow^a$ is not defined.

$$\underline{0 \uparrow^{x+iy} = 0 \quad \text{iff} \quad x > 0.}$$

(4) If $a \in \mathbf{Z}$, the restriction of $b$ to the cut plane is irrelevant and, in this case,

$$b \uparrow\!\!\uparrow^a = b \uparrow^a, \quad \text{and} \quad b^a \text{ is unambiguous.}$$

(5) $b \uparrow^a$ always denotes one definite number and so $p$, defined by $p(a, b) = b \uparrow^a$ for a suitable domain, is a *function*.

(6) For all $b$, $b \uparrow^{m/n}$ denotes one number; this value can only be selected *because* $b$ is restricted to the cut plane.
If $b \in \mathbf{R}$, $b > 0$, then $b \uparrow^{m/n} > 0$ because $b \uparrow^{m/n}$ denotes the real *positive* $n$th root of $b^m$. The interpretation of $r^{m/n}$ as $r \uparrow^{m/n}$ is almost universal when $r = |z|$.

(7) If $b \in \mathbf{R}$, $b \geq 0$,

$$b \uparrow^{1/2} = \sqrt{b}.$$

Replacement of (for example) $\sqrt{x}$ by $x^{1/2}$ before derivation or integration is equivalent to using $x^{1/2}$ to mean $x \uparrow^{1/2}$.

(8) When $a \in \mathbf{R}$, $b \in \mathbf{R}$, $b > 0$, $b^a$ *normally* means $b \uparrow^a$.
When convergent, the sum of the real binomial series is $(1 + x) \uparrow^a$ in all cases and when written $(1 + x)^a$, $(1 + x)^a$ is being used to mean $(1 + x) \uparrow^a$.

(9) Similarly, when the sum of the complex binomial series is written $(1 + z)^a$, $(1 + z)^a$ is being used to mean $(1 + z) \uparrow^a$.

(10) $\exp z$ coincides with $e \uparrow^z$.
cis $t$ coincides with $e \uparrow^{it}$.

**Table 9.2**

*Use of $b^a$ to mean $b \uparrow a$.*

(1) $b \uparrow a$ can be defined for all $a \in \mathbf{C}$ provided $b \neq 0$.

(2) If $b \neq 0$,

$$b \uparrow a = \exp(a \operatorname{Ln} b).$$

(3) The notation can, however, be extended to include $b = 0$ provided that the $x$-axal part of $a$ is then *positive*; otherwise, $0 \uparrow a$ is not defined.

$$\underline{0 \uparrow}^{x+iy} = 0 \quad \text{iff} \quad x > 0.$$

(4) If $a \in \mathbf{Z}$,

$$b \uparrow a = b \uparrow a, \quad \text{and} \quad b^a \text{ is unambiguous.}$$

(5) In general, $b \uparrow a$ has an unbounded number of values and does *not* define a function.
   If $a \in \mathbf{Q}$, the following *relations* are equivalent.

$$b \uparrow m/n = w \quad \Leftrightarrow \quad w^n = b^m.$$

(6) If $b \neq 0$, $b \uparrow m/n$ belongs to a set of $n$ numbers and, because $b$ is not restricted to the cut plane, a distinguished value can *not* be selected.
   If $b = \rho \operatorname{cis} \phi$,

$$b \uparrow m/n = \rho \uparrow m/n \operatorname{cis}\left(\frac{m}{n} \phi + \frac{2k\pi}{n}\right) \quad \text{for} \quad k = 0, 1, \ldots, n - 1.$$

[Note the modulus!]

(7) If $b \in \mathbf{R}$, $b \geq 0$,

$$b \uparrow 1/2 = \pm \sqrt{b}.$$

(8) Even if $a \in \mathbf{R}$, $b \in \mathbf{R}$, $b > 0$, $b^a$ *may* be used to mean $b \uparrow a$, particularly when it occurs as $b^{a+0i}$ or as $(b + 0i)^a$. [Ambiguity in such cases.]

(9) *Except* in connexion with the binomial series, $(1 + z)^a$ *may* mean $(1 + z) \uparrow a$. [Ambiguity in this case.]

(10) $e^z$ means $e \uparrow z$.
   $e^{it}$ means $e \uparrow it$.

*the negative real axis.* (In certain circumstances, $b$ *can* be allowed to be zero; see Table 9.1). The other symbol $b \mathbin{\text{⇑}} a$ will be used when $b$ is *not* restricted in that way and will denote $b^a$ on the occasions when it belongs to a *set* of values, $\exp(a \operatorname{Ln} b)$. The definitions and some illustrations of occurrences of the two interpretations are presented in Tables 9.1 and 9.2 (pages 136 and 137).

The notation $b \uparrow^a$ has been borrowed from computer programming. Since many of the influences of computer programming on pure mathematics have been harmful, the appropriation of one item from its vocabulary may be some compensation! Some computer languages already use $b \uparrow a$ to denote $b^a$ (the contexts always being ones in which the interpretation $b \uparrow^a$ rather than $b \mathbin{\text{⇑}}^a$ is intended). There would, however, be obvious advantages in printing and writing the index $a$ in its standard superscript position and *not* letting it sag to typewriter level except when actually typing.

If a notation something like this were ever adopted, suitable readings for $\uparrow$ and $\text{⇑}$ would have to be devised and the status of the traditional $b^a$ would have to be sorted out. It would not have to be abandoned altogether, because $b^a$ is unambiguous when $a \in \mathbf{Z}$ and so our deeply engrained habit of using $x^n$ in the writing of polynomials and power series would not be challenged. That leaves 4 possibilities. (1) Keep the symbol $b^a$ for use *only* when $a \in \mathbf{Z}$ and otherwise *always* write either $b \uparrow^a$ or $b \mathbin{\text{⇑}}^a$. Thus $16 \uparrow^{3/4} = 8$, while $16 \mathbin{\text{⇑}}^{3/4} = \pm 8$ when the algebra is real and $16 \mathbin{\text{⇑}}^{3/4} \in \{\pm 8, \pm 8i\}$ when it is complex. (2) Use $b^a$ in schools in the early stages (even for $b^{3/4}$ and so on), but only for as long as $b$ is restricted to be real and positive and $b^a$ always means $b \uparrow^a$. Abandon it as soon as the possibility of an ambiguous interpretation arises and write either $b \uparrow^a$ or $b \mathbin{\text{⇑}}^a$ from then on. [This, however, would involve unnecessary reteaching and could therefore be undesirable for the reasons mentioned in the Introduction.] (3) Keep $b^a$ as an alternative to $b \uparrow^a$, *never* using it for $b \mathbin{\text{⇑}}^a$, so that $b \uparrow^a$ would only replace $b^a$ when special emphasis was required (and the present $\zeta^{m/n}$ would *always* be written $\zeta \mathbin{\text{⇑}}^{m/n}$). The redundancy this causes would be unwelcome to pure mathematicians, although the optional infix symbol would appeal to computer programmers. (4) The worst solution. If $a \notin \mathbf{Z}$, keep $b^a$ as an imprecise alternative (as now) for *either* $b \uparrow^a$ *or* $b \mathbin{\text{⇑}}^a$ but, at the beginning of any piece of work in which the same interpretation is to be used throughout, specify which meaning, $b \uparrow^a$ or $b \mathbin{\text{⇑}}^a$, the shorthand $b^a$ will be given.

For the present, however, all that aspiring teachers can do is to make themselves familiar with the issues involved, so that they are prepared for the classroom confusion the traditional notation generates, even when the algebra is real.

# Chapter 10

# Derivation, differentiation and integration

## 10.1

Let us first get our language clear. Here, as so often happens, mathematicians are not consistent. The pupils says: 'When I differentiate $3x^2 + 2x$, I get the differential $6x + 2$'. 'Differential' is the wrong word: 'differential' is *not* synonymous with 'derivative'. But the student, unlike the mathematician, is being quite logical. He is permitted to say: 'When I integrate $3x^2 + 2x$, I get the integral $x^3 + x^2 + A$'. Since, when he performs the *operation* of integration, the *result* he obtains is called an integral, he assumes, not unreasonably, that, when he performs the operation of differentiation, his result can be called a differential. Alas, however, that is not so and it is rather too late now to change the meaning of the word 'differential', which describes an entity like $dy$ or $dx$ when it is used independently. The only redeeming feature of this unfortunate muddle is that the majority of pupils learning calculus will perhaps not need to meet the concept of a differential at all while they are still at school. So, if they never read or hear the word 'differential' used as a noun, they may not be tempted to invent such a use for themselves! The word 'differential' has, of course, also an adjectival use, as in 'differential equation', 'differential geometry', etc.

The word to describe $dy/dx$ or $f'(x)$ is *derivative* and $f'$ is called the *derived function*. The distinction between these two is not always as rigidly maintained as it ought to be. This is a legacy from that earlier era when functions were less well understood and, as discussed in Chapter 2, $g(x)$ and $g$ were not properly distinguished. Note also that the not uncommon practice of mixing the two notations and writing $df/dx$ is, logically speaking, quite indefensible, whichever of these two concepts this hybrid symbol is deemed to denote: please pause and think about this.

The confusion of names has arisen because of the once-fashionable description of the derivative as a 'differential coefficient', which careless and lazy people have later truncated back to 'differential'. This appellation was possible because, following an overhasty introduction of the idea of a differential, $dy$ was written as

$$dy = f'(x)\,dx \quad \text{or} \quad dy = \frac{dy}{dx}\,dx,$$

so that the derivative, $f'(x)$ or $dy/dx$, could be thought of as the coefficient of one differential, $dx$, in the expression for the other, $dy$. Nowadays, any study of differentials, as opposed to derivatives, comes much later in the young mathematician's education so that the rationale for that curious phrase 'differential coefficient' is lacking. Fortunately, however, the term 'differential coefficient' is now obsolete—or very nearly so.

One or two writers have devised a brilliant solution to this dilemma, but their terminology is not yet widely known, much less adopted. The word *'derivation'* is introduced to describe the operation of getting from $y$ to its derivative, $dy/dx$. The word 'differentiation' is not used at all until the stage when differentials are eventually defined and it then describes the operation of getting from $y$ to its differential $dy$. So, if you like, we can perhaps stretch our mapping notation a little and sum this nomenclature up succinctly by writing

$$\text{derivation: } y \mapsto \frac{dy}{dx}; \quad \text{differentiation: } y \mapsto dy.$$

Thus, one *derives* $y = x^4 - 3x^2$ (with respect to $x$) to get the *derivative*

$$\frac{dy}{dx} = 4x^3 - 6x,$$

but one *differentiates* $y = x^4 - 3x^2$ to get the *differential*

$$dy = (4x^3 - 6x)\,dx.$$

This distinction[†] becomes even more valuable when applied to

---

[†]The two operations were carefully distinguished in, for example, de la Vallée Poussin's *Cours d'analyse infinitésimal*, but in many later books they have not been. One notable elementary English textbook in which this distinction is maintained with particular clarity is F. Gerrish *Pure mathematics* (C.U.P.).

functions of more than one variable, when it becomes a distinction between *partial derivation* and *differentiation*. (The familiar term 'partial differentiation' has no place in this scheme.)

Now, one *derives* $w = x^3y^5z$ (wo $x, y, z$) to get the (*partial*) *derivatives*

$$\frac{\partial w}{\partial x} = 3x^2y^5z, \qquad \frac{\partial w}{\partial y} = 5x^3y^4z, \qquad \frac{\partial w}{\partial z} = x^3y^5;$$

one *differentiates* $w = x^3y^5z$ to get the *differential*

$$dw = 3x^2y^5z \, dx + 5x^3y^4z \, dy + x^3y^5 \, dz.$$

Once one has overcome one's surprise at finding the word 'derive' used in this way, the advantages of the terminology become clear (although the new usage of 'derive' may perhaps be held to conflict slightly with the other mathematical sense of the word, when a conclusion is derived from a hypothesis). You will probably not feel sufficiently confident to abandon such a familiar word as 'differentiate' and replace it with 'derive' until you have gained some experience: indeed, your colleagues in both mathematics and science will certainly not thank you if you do so without their approval! But the writer will try to make the words 'derivation' and 'derive' more familiar by using them in this book and you will then be better able to judge whether you would find them acceptable or not. The distinction just described between these words and the terms 'differentiation' and 'differentiate' will be maintained.

This punctiliousness yields a further bonus when one comes to functions of two or more variables. When, in your analysis courses, you first met the concept of 'differentiability' for such functions, you may well have found the definition strange and unexpected, because the property turned out to involve greater restriction on the function than was needed to ensure the mere existence of all the first-order partial derivatives. Differentiability is, in fact, a quite natural idea and, in the next chapter, the motivation for the definition is examined.

The point to be made here, however, is that the word 'derivable' is now available and can be applied in a natural way to a function all of whose first derivatives exist (at a particular point). This provides a useful counterpoint to the term 'differentiable', which continues to describe the other, more subtle property. When its association with the operation of differentiation has been properly investigated, it will become clear that, whereas

*derivable* (at a point) means that    *derivation*  is possible there,
*differentiable* (at a point) means that *differentiation* is possible there.

It will also emerge that

differentiability      ⇒      derivability

but *not* conversely, in general. The new terminology is not only clearer, it is more complete and consistent, since it has a comprehensive and well-defined set of terms for each of the two dissimilar ideas.

With functions of a *single* real variable, however, it happens that the concepts of derivability and differentiability are equivalent, which is, of course, why they can become confused in people's minds. The one that is needed at school is that of *derivability:* $f$ is *derivable* at $a$ if a derivative $f'(a)$ exists, that is, if $[f(a + h) - f(a)]/h$ tends to a limit as $h \to 0$. The other notion, differentiability, will not be studied until much later. It is examined in Chapter 11 and there shown to be equivalent to derivability in the case of a function of one variable, but not in general.

### 10.2

There are two ways of illustrating the derivative graphically: (1) as a gradient, (2) as a scale factor. Inevitably, many student teachers will only have experienced one of these two treatments themselves and may be apprehensive about the prospect of having to teach the other. Actually, there is less difference than there appears to be. What has to be realized is that a *gradient* is merely the natural interpretation of a derivative when the original function is pictured by means of a *cartesian* graph† and a *scale factor* is the appropriate interpretation when the function is represented by a *mapping* graph. Moreover, one must recognize that neither of these graphic expedients really goes to the heart of the concept of the derivative d$y$/d$x$ as the *rate of change* of $y$ with respect to $x$. This is the notion of paramount importance, which must be got across whichever pictorial illustration is used.

Both expositions start in the same way, by letting $x$ increase by $h$ (or, at a later stage, $\delta x$) from $a$ to $a + h$. The increase this produces in $y$ is then denoted by $k$ (or $\delta y$). The variation now comes in the graphic interpretation given to $k/h$ (or $\delta y/\delta x$). In method (1), $k/h$ is seen to be the *gradient of the chord* PQ in the familiar cartesian graph in Fig. 10.1. In method (2), $k/h$ is the magnification factor of the interval $[a, a + h]$, or the *average scale factor* over this interval, an idea that is intimately

---

†As explained in Section 2.18, the cartesian graph, as conventionally drawn, is, strictly speaking, the graph of the *relation* $y = f(x)$, rather than of the *function f*. The remarks in Section 2.17 about the notation used for the dependent variable are also very relevant here.

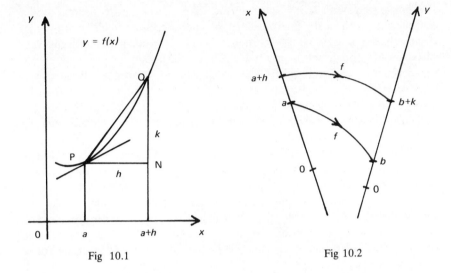

Fig 10.1                                    Fig 10.2

associated with the mapping graph in Fig. 10.2, but is extraneous to the cartesian plot.

If $f$ is derivable at $a$, then, as $h \to 0$, the gradient of the chord PQ in Fig. 10.1 tends to the *gradient of the tangent* to the cartesian graph at P, whereas in Fig. 10.2 the average scale factor over the interval $[a, a + h]$ tends to the *local scale factor* at $a$. In both treatments, it is, of course, vital to consider negative as well as positive $h$ to establish that the limits as $h \nearrow 0$ and $h \searrow 0$ exist and are the same, otherwise it is not legitimate to talk about the limit as $h \to 0$ or to write 'lim'. [See Section 5.1.]
$$\underset{h \to 0}{}$$
The next stage is to look at the limiting process more analytically and less graphically, emphasizing that an increase $h$ in $x$ has produced an increase $k$ in $y$, so that $k/h$ measures the *average rate of increase* of $y$ with respect to $x$ in the interval $[a, a + h]$. After the usual agreement that the *rate of increase* of $y$ with respect to $x$ at $a$ will be *defined* as the limit of $k/h$ as $h \to 0$, the interpretation of the derivative as this rate of change is achieved. In so far as pictorial arguments are needed to reinforce this, neither graph has any particular advantage over the other.

What, then, are the relative merits of the two graphic methods? Method (2) is useful for strengthening the important idea of a function as a mapping, since the derived function also becomes associated with a particular feature of the mapping graph. The scale factor idea also wins hands down in dealing with the 'chain rule' for derivation,

$$\frac{dy}{du} = \frac{dy}{dx} \cdot \frac{dx}{du}.$$

If a mapping $f$, with scale factor 3 at $a$, maps $a$ to $b$ and is followed by a mapping $g$, with scale factor $-2$ at $b$, then the composite mapping $gf$ clearly has a scale factor $-6$ at $a$: the chain rule is made intuitively obvious. On the other hand, the value of the mapping graph does not mean that the cartesian graph should or can be ignored—it is much too important—and, certainly, the gradient of the tangent is often felt to be a more tangible realization (no pun intended) of the derivative than the scale factor. Other points in favour of the cartesian graph will emerge later.

It is sometimes argued that the calculation of speed by derivation of distance (and similar kinematic problems) is more closely allied to the scale factor treatment, but this is not so. No mapping from time ($t$) to distance[†] ($x$) need be explicitly introduced. Of course, such a mapping *can* be defined, but then, if one wants to, a cartesian distance–time graph can also be drawn: neither, however, is essential. One merely has to observe that a distance $\delta x$ is covered in time $\delta t$, giving an average speed

| time | | $t$ | | $t + \delta t$ |
|------|--|-----|--|----------------|
| | | A | | B |
| distance | | $x$ | | $x + \delta x$ |

$\delta x/\delta t$ in the interval from A to B. Then, with the usual definition of the speed *at* time $t$ (namely, $\lim\limits_{\delta t \to 0} \delta x/\delta t$), the association of $dx/dt$ with the speed at time $t$ is obtained.

## 10.3

A much clearer understanding of the two aspects of the derivative can be acquired by looking at functions of more than one real variable. This observation will be no surprise to a true mathematician: very often a more general formulation of a problem enables its essential features to be clearly perceived.

If $w = f(x, y, z)$ and $f$ is derivable,

$$\frac{\partial w}{\partial x} = \lim_{\delta x \to 0} \frac{f(x + \delta x, y, z) - f(x, y, z)}{\delta x}$$

and so on. The partial derivatives are still rates of change of $w$ with respect to one other variable, even though these rates have to be

---

[†]Since $x$ is an algebraic scalar and not an unsigned number, it would be preferable to call $x$ a position coordinate rather than a distance, but this sloppy usage has become hallowed by custom.

calculated under the special condition that all the remaining variables are held constant while the limiting process takes place.

Notice, in passing, that the preferred alternative notation for the above partial derivative is $f_1(x, y, z)$ and *not* $f_x(x, y, z)$, despite the popularity of the latter notation among our scientific colleagues. Some analysts become quite apoplectic if $f_x$, $f_y$, $f_z$ are used for the derived functions instead of $f_1$, $f_2$, $f_3$. There is no gainsaying the validity of their objection: the notation $f_x$ for a derived function is logically just as indefensible as is the use of $\partial f/\partial x$ for a partial derivative. Also, although $f_1$ denotes a clearly defined mapping, the notation $f_x$ lends itself to misinterpretation in many contexts: for example, when $f(x, y, z)$ and $f(y, x, z)$ both occur, or when $w = f(x, y, z)$ where $z = g(x, y)$. Nevertheless, the widespread use of the abhorred alternative notation by scientists cannot—unfortunately—be ignored.

In point of fact, $f_1$, $f_2$, $f_3$ is not a particularly inspired notation. For *well-behaved* functions, a more systematic choice would be $f^{(100)}$, $f^{(010)}$, $f^{(001)}$, which not only represents a natural evolution of the notation used with functions of one variable, $f'$ or $f^{(1)}$, but is capable of smooth generalization to $f^{(lmn)}$ for all higher order derived functions.

If $w$ is a function of just 2 variables, the cartesian graph of $w = f(x, y)$ is a surface in 3-dimensional euclidean space. By keeping $y$ constant, the increases in $x$ and $w$ are confined to a plane perpendicular to the $y$-axis and $\partial w/\partial x$ measures the gradient in that plane of the curve of intersection of the plane and the surface. So $\partial w/\partial x$ and $\partial w/\partial y$ both measure gradients associated with the cartesian graph, albeit of sections of that graph. If $w$ is a function of more than 2 variables, the cartesian graph requires more than 3 euclidean dimensions, which, of course, makes the representation less practically useful. But theoretically the partial derivatives still have exactly similar geometric meanings.

As a mapping, $f$ now maps the plane $\mathbf{R}^2$ (or, more generally, the space $\mathbf{R}^n$) to a line (Fig. 10.3). Although, as you will see, there *are* interpretations of $\partial w/\partial x$ and $\partial w/\partial y$ as scale factors associated with certain directions through the point $(x, y)$, there is no longer any one feature of the mapping itself that is felt to correspond very closely to the

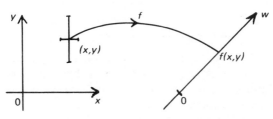

Fig 10.3

idea of a 'local' scale factor at a point of the domain, which was what provided the illustration for $dy/dx$ when $y = f(x)$.

Moreover, the chain rule for derivation is lost with partial derivatives. If $w = f(x, y)$, where $x = g(u, v)$, $y = h(u, v)$, and the functions $f, g, h$ are all derivable, then, instead of the nice simple chain rule, one has the more complicated formulae

$$\frac{\partial w}{\partial u} = \frac{\partial w}{\partial x} \frac{\partial x}{\partial u} + \frac{\partial w}{\partial y} \frac{\partial y}{\partial u}$$

$$\frac{\partial w}{\partial v} = \frac{\partial w}{\partial x} \frac{\partial x}{\partial v} + \frac{\partial w}{\partial y} \frac{\partial y}{\partial v}.$$

Paradise has been lost.

Written in matrix form, however, these formulae become

$$\begin{bmatrix} \dfrac{\partial w}{\partial u} \\ \dfrac{\partial w}{\partial v} \end{bmatrix} = J \begin{bmatrix} \dfrac{\partial w}{\partial x} \\ \dfrac{\partial w}{\partial y} \end{bmatrix} \quad \text{where} \quad J = \begin{bmatrix} \dfrac{\partial x}{\partial u} & \dfrac{\partial y}{\partial u} \\ \dfrac{\partial x}{\partial v} & \dfrac{\partial y}{\partial v} \end{bmatrix},$$

so that J depends only on the mapping $(u, v) \mapsto (x, y)$ and not on the function $f$.

You will recall that the determinant of this matrix J is called the *jacobian* of the mapping, written $\partial(x, y)/\partial(u, v)$, so that

$$\frac{\partial(x, y)}{\partial(u, v)} = \begin{vmatrix} \dfrac{\partial x}{\partial u} & \dfrac{\partial y}{\partial u} \\ \dfrac{\partial x}{\partial v} & \dfrac{\partial y}{\partial v} \end{vmatrix}$$

Moreover, the jacobian has an important property in relation to the mapping. If $(U, V) \mapsto (X, Y)$ and a region S of area $\sigma$ containing $(U, V)$ has (see Fig. 10.4) image T of area $\tau$ (necessarily containing $(X, Y)$), then the superficial magnification factor at the point $(U, V)$, that is, $\lim_{\sigma \to 0} (\tau/\sigma)$, is equal to the absolute value of the jacobian, $| \partial(x, y)/\partial(u, v) |$, evaluated at $(U, V)$. (The significance of the sign of the jacobian is interesting, and was mentioned in Section 7.6, but is not important for the present purpose.) In other words, *the jacobian is a local scale factor*, although it is a superficial and not a linear one.

Also, it is well known, and quite easily proved, that, unlike partial

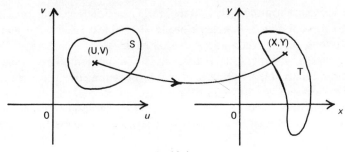

Fig 10.4

derivatives, jacobians *do* enjoy a chain rule property, namely

$$\frac{\partial(x, y)}{\partial(u, v)} = \frac{\partial(x, y)}{\partial(\xi, \eta)} \cdot \frac{\partial(\xi, \eta)}{\partial(u, v)} ,$$

which, of course, is no surprise once their interpretation as scale factors is recognized. Paradise has been regained.

The extensions to $\partial(x, y, z)/\partial(\xi, \eta, \zeta)$ and so on are quite straight-forward.

So, to summarize, the gradient aspect of the (ordinary) derivative extends in a fairly natural way to partial derivatives, but the analogue of the scale factor property is enshrined in the jacobian and *not* in any individual partial derivative. Those who prefer to teach ordinary derivatives using scale factors should keep this fact firmly in mind.

Note also that rate of change, in the sense of a rate of change of one variable with respect to another, which is the usual sense, attaches itself (along with the gradient property) to the partial derivatives. The scale factor is *not* now associated with this sort of rate of change, only with the less useful idea of the rate of change of an area in one plane with respect to an area in another (or, of course, the generalization of this idea to three or more dimensions).

Turning these observations the other way round, one may say that partial derivatives and jacobians embody two different aspects of certain functional relations. With functions of a single real variable, *both these features coalesce* in the (ordinary) derivative, which performs both roles. Thus the existence of two aspects of this derivative is not really any cause for astonishment, but the consequences of this ambivalence for later extensions are important.

## 10.4

It is hoped that the above discussion will have helped to put the two associated graphic treatments into better focus. Readers who are knowledgeable about complex analysis may be interested to take a brief look at a function of a (single) complex variable to see what geometric picture of the derivative emerges in that case.

Let $w = f(z)$ for all $z$ belonging to some suitable (and clearly defined) domain D. A cartesian representation of a function of a complex variable would require the use of 4 real dimensions, so a mapping diagram becomes the only practical graphic device available and in this subject it is, of course, exploited with tremendous effect.

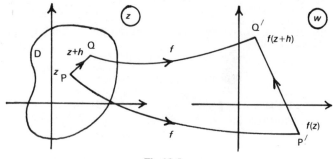

Fig 10.5

You will recall that $f$ is derivable at $z$ if

$$\frac{f(z + h) - f(z)}{h},$$

(where $h$ also is complex), tends to a limit as $h \to 0$ (in any manner) and, if so, this limit is denoted by $f'(z)$. With the notation in Fig. 10.5, $f(z + h) - f(z)$ is represented by $\overrightarrow{P'Q'}$ and $h$ by $\overrightarrow{PQ}$. So, in magnitude, $P'Q' : PQ$ tends to $|f'(z)|$ as $h \to 0$ and $|f'(z)|$ is again a local scale factor: the linear magnification factor at P. Note incidentally that the enlargement is locally isotropic, that is, the same for all directions through P.

Also, taking the axes in the $z$ and $w$ planes to be parallel, and assuming $f'(z) \neq 0$, $\arg f'(z)$ measures the limiting value of the rotation from $\overrightarrow{PQ}$ to $\overrightarrow{P'Q'}$ as $h \to 0$. A trifling refinement of this reasoning establishes that the mapping $f$ is *conformal* at all points $z$ at which $f'(z) \neq 0$. This means that if 2 curves $C_1$ and $C_2$, intersecting at P, have images $C_1'$ and $C_2'$ under $f$, intersecting at P', then the rotation at P from $C_1$ to $C_2$ is equal in magnitude and sense to that at P' from $C_1'$ to $C_2'$.

So, both the modulus and the argument of $f'(z)$ have geometric significance for the mapping and $|f'(z)|$ is a scale factor.

**10.5**

There are some other observations about the notation for derivation and integration that are worth making. It is fashionable (even trendy) to write $\int f = F$, concentrating on the *integrated function*, F, rather than the *integral*, $F(x)$, just as one can write $F' = f$, concentrating on the *derived function*, f, rather than the *derivative*, $f(x)$. This is perfectly satisfactory, even admirable, provided the limitations of the notations are recognized and strictly adhered to. As so often happens, however, not all followers of the latest vogue have troubled to find out whether they understand it or to think through the consequences of their enthusiasm.

The statements

$$\frac{d}{dx}(x^3) = 3x^2, \qquad \frac{d}{dy}(y^3) = 3y^2, \qquad \frac{d}{dp}(p^3) = 3p^2$$

are obviously all expressing the same mathematical fact. Derivation, viewed in the most fundamental way, is not an operation that is applied to an expression to give a derivative; it is an operation that produces from a given function another function (the derived function). If the mapping (with domain **R**)

$$x \mapsto x^3, \qquad y \mapsto y^3, \qquad p \mapsto p^3$$

(these are all the same mapping) is denoted by F and the mapping

$$x \mapsto 3x^2, \qquad y \mapsto 3y^2, \qquad p \mapsto 3p^2$$

by f, the above statements all become

$$F' = f,$$

without any mention of the extraneous and distracting features involving $x, y$ and $p$.

To this extent, the new statement is clearer, more concise and more fundamental. But, of course, the price paid for this simplicity is that one has to remember all the time what mappings F and f are representing, whereas the conventional statements with expressions rather than functions convey the pattern clearly.

With functions symbolized by prefix notations, one fares a little better. One *can* write

$$\sin' = \cos, \quad \cos' = -\sin, \quad \exp' = \exp$$

without any qualms (although the wisdom of doing this with pupils is debatable). But one cannot write, for example,

$$\sec' = \sec \tan,$$

because $\sec \tan (x)$ means $\sec(\tan x)$ and *not* $\sec x \tan x$. Similarly,

$$\tan' = \sec^2$$

is wrong because, in this context, $\sec^2(x)$ would *have* to mean $\sec(\sec x)$ and *not* $(\sec x)^2$. [See Section 6.4 on the illogicality of the $\sin^2$ notation.] Nor can one write

$$\ln' = \frac{1}{x} \qquad (x > 0),$$

since $1/x$ is a real number and not a mapping; one would have to define a reciprocal mapping

$$S \to S : \lambda : x \mapsto \frac{1}{x}$$

where S is the set of positive real numbers, in order to write

$$\ln' = \lambda$$

and it cannot be said that anything has been gained in clarity or insight by this manoeuvre.

Add to all this the fact that most functions are not symbolized by prefix notations without special definitions being made and it will be seen that the abbreviated statement $F' = f$ is really only useful for theoretical purposes in analysis when general functions are being discussed, and not for practical examples. Is the bonus, then, sufficient for the use of this notation by pupils at school to be advocated?

### 10.6

The same considerations apply just as forcefully to integration, but here there is a further complication. When integrating $\sin x$, the statement one hopes the student will make is

$$\int \sin x \, dx = -\cos x + A;$$

the trendy alternative

$$\int \sin = -\cos$$

does not carry the same precision and, of course,

$$\int \sin = -\cos + A$$

is gibberish. On the other hand,

$$\int_{\pi/4}^{\pi/2} \sin = \frac{1}{\sqrt{2}}$$

is acceptable, at least in principle.

If one is given the mapping

$$f : x \mapsto 3x^2$$

and wishes to write $F = \int f$, one cannot just define

$$F : x \mapsto x^3 + A$$

because, with the arbitrary constant floating about, this does not define a mapping—at least, not unless more care is taken to specify when the choice of A is to be made. [The definition of a mapping is discussed in Section 2.6.] Even if this difficulty with the constant is glossed over and statements like

$$\int ch = sh$$

are accepted as useful, the number of functions for which this is going to be possible without copious extra symbol definition is insignificant. Certainly, for example, a claim like

$$\int \sin \cos = \tfrac{1}{2} \sin^2$$

is inadmissible, for the sorts of reason mentioned earlier: $\sin \cos (x)$ means $\sin(\cos x)$ and $\tfrac{1}{2}\sin^2(x)$ means $\tfrac{1}{2}\sin(\sin x)$, *not* what they would have to mean for the claim to make sense. Nor can one mix the two notations and try to write

$$\int \frac{1}{1 + x^2} = \tan^{-1};$$

one would have to define a mapping

$$\mathbf{R} \to \mathbf{R} : \mu : x \mapsto \frac{1}{1 + x^2}$$

in order to say

$$\int \mu = \tan^{-1}:$$

so the hopes of the lazy pupil that he is at last going to be allowed to miss off those 'd$x$' that he finds so tiresome are *not* going to be fulfilled.

Note also that confusing the two legitimate statements $\int f = F$ and $\int f(x) \, dx = F(x)$ by writing $\int f \, dx$ is really just as blameworthy as writing $\int f(x)$, missing off the 'd$x$'. It is analogous to mixing the permissible notations $f'$ and $dy/dx$ by writing $df/dx$.

More interestingly, the writer has seen

$$\int \sin 2 = -\tfrac{1}{2} \cos 2$$

used (for $\int \sin 2x \, dx = -\tfrac{1}{2} \cos 2x + A$). One must reluctantly admit that (provided one ignores the problem with the arbitrary constant A) it is difficult to object to this bizarre statement on logical grounds, but the opportunities it opens up for pupil misunderstanding are too dire to contemplate.

While accepting the elegance and economy of statements like $\int f = F$ for theoretical work, and approving the emphasis they put on the mapping concept, it is difficult to share the enthusiasm of those who hope to encourage the spread of such notation to the schools.

Chapter **11**

# Differentials

---

## 11.1

In the previous chapter it was suggested that, with the majority of pupils, if not all, it is probably best not to introduce the topic of differentials at all. But this is another area where liaison with one's science colleagues is necessary, since they may be anxious to use them. Whether they are or not, you yourself may be keen to try teaching this subject to some of your brighter students and so some discussion of the problems it poses will perhaps not be amiss. Certainly, if you are planning to teach partial derivation, there are considerable advantages to be gained by introducing differentials, although the initial expenditure of effort is far from negligible, if you are really going to achieve understanding and not just meaningless symbol-crunching.

Let us first be quite clear on one point, however. If your reason for wishing to teach differentials is that you imagine there is some payoff in integration theory or solving differential equations, then forget it: you are going to do it for completely the wrong reason. Even when the definition of the differential $dx$ has been made, that does *not* give any independent meaning to the $dx$ that occurs in $\int \ldots dx$. *That* $dx$ is a purely formal part of the symbol $\int \ldots dx$, without any life of its own. It is not, strictly speaking, a differential at all. Of course, the rules that govern the change of variable in an integral make the notation $\int \ldots dx$ extremely *suggestive* for getting the right answer, which is precisely why that notation has been adopted. It is most important to appreciate that, even when the differential $dx$ has been satisfactorily defined (in a way that will be discussed presently), that does *not* give any new and alternative way of defining $\int f(x)\,dx$ or proving theorems about integration.

**11.2**

The Riemann integral $\int_a^b f(x)\,dx$ is defined as a limit of a sum. A dissection

$$a = x_0 < x_1 < x_2 < \ldots < x_{n-1} < x_n = b$$

is taken of the interval $[a, b]$. The length of the subinterval $[x_{r-1}, x_r]$ is denoted by $h_r$ $(1 \leqslant r \leqslant n)$ and the length of the largest subinterval by H; that is,

$$h_r = x_r - x_{r-1}, \qquad H = \max_r h_r.$$

Then, for each $r$, $\xi_r \in [x_{r-1}, x_r]$ is taken and the sum $\sum_{r=1}^n f(\xi_r)h_r$ is constructed. If and only if this sum tends to a limit as $H \searrow 0$, then $f$ is said to be (Riemann) integrable in $[a, b]$ and the symbol $\int_a^b f(x)\,dx$ is used to denote this limiting sum; that is,

$$\int_a^b f(x)\,dx = \lim_{H \searrow 0} \sum_{r=1}^n f(\xi_r)h_r.$$

This is only a sketchy summary of an intricate process, but it should be sufficient to remind you of theory you will have studied in your mathematics courses. It should persuade you that the $dx$ in the definite integral serves a merely formal purpose, with no independent existence.

**11.3**

What about the *indefinite* integral? This is introduced in response to the challenge of finding a 'primitive' or 'antiderivative': the statement

$$y = \int f(x)\,dx$$

in which an indefinite integral occurs is presented as just another way of writing

$$\frac{dy}{dx} = f(x)$$

(just as $y = e^x$ is another way of writing $\ln y = x$).

But why, you may say—or, when they have met differentials, your pupils may say—can I not write this last statement in the form

$$dy = f(x)\,dx$$

and say that I am 'integrating both sides' to get

$$\int dy = \int f(x)\, dx$$

and hence

$$y = \int f(x)\, dx?$$

The true answer is that one cannot take any old expression and just 'integrate' it in some general and unspecified way[†] (in the same sort of way that one might take an expression and just 'double' it). All one can legitimately do is take an expression of an appropriate form and integrate it *with respect to a specified variable.*

Consider the method for changing a variable in an integral. If

$$y = \int f(x)\, dx \quad \text{and} \quad x = g(u),$$

then, of course, it turns out that

$$y = \int f[g(u)]\, g'(u)\, du.$$

Various fallacious justifications are offered for this, so let us start by considering what is really involved. Since

$$y = \int f(x)\, dx, \tag{1}$$

$$\frac{dy}{dx} = f(x).$$

By the chain rule,

$$\frac{dy}{du} = \frac{dy}{dx}\frac{dx}{du},$$

and so, since $x = g(u)$, this becomes

$$\frac{dy}{du} = f[g(u)]\, g'(u). \tag{2}$$

[†]There is a sense in which it would be possible to give a coherent presentation of integration on these lines *if* integration were defined as the operation inverse to differentiation, rather than as the operation inverse to derivation: if the indefinite integral were introduced as an 'antidifferential' instead of as an 'antiderivative'. But any such account would postpone the study of integration until after the introduction of differentials and would be unacceptable for school use anyway. It is assumed here that the traditional definitions and points of view have been adopted when pupils first studied integration.

Then, integrating *with respect to u,*

$$y = \int f[g(u)]\, g'(u)\, \mathrm{d}u, \tag{3}$$

which is the required formula.

Going from (2) to

$$\mathrm{d}y = f[g(u)]\, g'(u)\, \mathrm{d}u \tag{4}$$

and formally prefixing each side with a sign '$\int$' gives the right answer—the notation is successful precisely because this happens—but it is nonsense to say that (3) is obtained from (4) by 'integrating both sides' of (4).

Another (equivalent) fallacious justification is to say that, if $x = g(u)$, the differentials $\mathrm{d}x$ and $\mathrm{d}u$ satisfy

$$\mathrm{d}x = g'(u)\, \mathrm{d}u,$$

and so (3) is obtained from (1) by replacing $x$ by $g(u)$ and 'replacing the "differential" $\mathrm{d}x$ by $g'(u)\,\mathrm{d}u$, where $\mathrm{d}u$ is another "differential" '. It cannot be too strongly emphasized again that the $\mathrm{d}x$ and $\mathrm{d}u$ in $\int \ldots \mathrm{d}x$ and $\int \ldots \mathrm{d}u$ are *not* differentials. They are written in the same way as the differentials $\mathrm{d}x$ and $\mathrm{d}u$ *because*, as we can see, no confusion arises and to that extent the notation is visually very helpful and suggestive in getting correct results. But what may be admirable as a mnemonic cannot be inflated to serve as a proof. It is not differentials that prove the change of variable result: it is the change of variable result that *justifies the notation* $\int \ldots \mathrm{d}x$, with $\mathrm{d}x$ being written in the same way as a differential.

It is because one formally gets correct results by treating $\mathrm{d}x$ in $\int \ldots \mathrm{d}x$ *as though* it were a differential that this notation survives and continues to flourish, despite the battle every generation of teachers has to wage to prevent their pupils omitting the '$\mathrm{d}x$'. But that does not mean that one can abuse the notation by pretending that it provides a justification for the change of variable algorithm, or any similar manipulation.

[An analogy may commend itself: $b^0 . b^n$ is *not* equal to $b^n$ *because* the index laws insist that $b^m . b^n = b^{m+n}$, so that, when $m = 0$, $b^0 . b^n$ automatically equals $b^{0+n} = b^n$: they are equal because, looking at the index laws, one saw that, by *defining* $b^0 = 1$, one was able to make a useful definition that would be *consistent* with those laws.]

As a final example of these horrors, consider the solution of a first-order differential equation with separable variables, such as

$$\frac{dy}{dx} = \frac{p(x)}{q(y)} .$$

One often hears people say that the solution is obtained by 'crossmultiplying' (O word of ill omen!), giving

$$q(y) \, dy = p(x) \, dx,$$

and then 'integrating each side with respect to its own variable' (!), getting

$$\int q(y) \, dy = \int p(x) \, dx.$$

The fact that the right answer *is* obtained in this way is merely a tribute to the felicity of the notation. As an explanation of what is actually going on, it is nonsense.

All that is done, of course, is that the original differential equation is rewritten

$$q(y) \, \frac{dy}{dx} = p(x)$$

and each side is then integrated *with respect to x*—please note—giving

$$\int q(y) \, \frac{dy}{dx} \, dx = \int p(x) \, dx.$$

Then, applying the change of variable rule, the left side becomes $\int q(y) \, dy$, so that

$$\int q(y) \, dy = \int p(x) \, dx.$$

An alternative explanation would be to say that $q(y) \, dy/dx$ may be written $d/dx[r(y)]$, where $r' = q$, so that

$$p(x) = \frac{d}{dx} [r(y)]$$

and

$$\int p(x) \, dx = r(y)$$
$$= \int q(y) \, dy.$$

By now, we hope, readers will agree that differentials should not be taught in the hope that they will offer a magic recipe for simplifying integration or the solution of differential equations, despite the useful mnemonics they provide for rules whose proof depends on quite other considerations.

## 11.4

The important and genuine applications of differentials arise in connexion with partial derivation, but the starting point for the theory will be a function of a single real variable. Its cartesian, rather than its mapping, graph provides the appropriate diagram.

So often, the only attempt a teacher or writer makes at a definition of a differential is to say that, in Fig. 11.1, PN is denoted by $dx$ and NT by $dy$. Then, since the gradient of PT is $f'(x)$,

$$dy = f'(x)\, dx.$$

'So what?', says the student. Even the pivotal result that this relation continues to hold if the variable $x$ is not independent may not be alluded to.

The 'definition' offered is scarcely adequate and complete mystification will be produced when

$$dw = f_1(x, y, z)\, dx + f_2(x, y, z)\, dy + f_3(x, y, z)\, dz$$

is conjured out of the hat as the 'natural generalization' of this formula to a function of more than one variable. If the topic is important enough to be taught at all, then surely it deserves a more coherent presentation than this.

What is *given* is a function $f : x \mapsto y$, a value $x$ belonging to the domain[†] of $f$ and an increment $\delta x$ ($\neq 0$) in that variable. The differential $dy$ is then *defined* by

$$dy = f'(x)\, \delta x. \tag{5}$$

Notice particularly (a) that $dy$ is a function of *two* variables $x$ and $\delta x$ and (b) that there is no mention (yet) of $dx$. This definition is chosen so

---

[†]Here and elsewhere in this chapter, the domains of all functions mentioned are being assumed (for technical reasons) to be *open* sets, even when this is not explicitly stated. This ensures that, for all $x \in D \subseteq \mathbf{R}$, there is some $\eta$ such that $x + \delta x$ *also* belongs to D for all $|\delta x| < \eta$, with the corresponding extensions of this property when the domains are more than 1-dimensional.

that, for the *given f*, the *given x* and the *given δx*, the differential d*y*
measures in Fig. 11.1 the step $\overline{\text{NT}}$ to get from N (at the same horizontal
level as P) up to T (on the tangent at P); just as δ*y*, defined as
$f(x + δx) - f(x)$, measures the step $\overline{\text{NQ}}$ to get from N up to Q (on the
curve). [The bar has been added to remind you that the step is a *signed*
length $(\overline{\text{TN}} = -\overline{\text{NT}})$. There is no standard notation to emphasize this. It
is tempting to use an arrow $(\overrightarrow{\text{NT}})$ to suggest that it has some similarity
to a vector, but that would be misguided, since these signed lengths are
presently going to be divided, and such division, of course, would not be
possible if they really were vectors.]

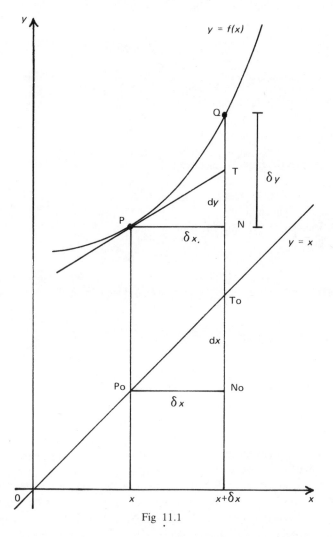

Fig 11.1

So, if $y = x^5$,

$$dy = d(x^5) = 5x^4 \, \delta x;$$

if $y = 1/x^2$, $x = 4$, $\delta x = 0.8$,

$$d\left(\frac{1}{x^2}\right) = -\frac{2}{4^3} \times 0.8 = -0.025;$$

if $f(x) = \cos x$, $x = \frac{1}{3}\pi$, $\delta x = -1.5$,

$$d(\cos x) = -\frac{\sqrt{3}}{2} \times (-1.5) \simeq 1.299,$$

and so on.

In particular, if $f(x) = x$,

$$d(x) = 1 \times \delta x$$

and so, writing $dx$ rather than $d(x)$,

$$dx = \delta x. \tag{6}$$

Substituting (6) in (5), the formula

$$dy = f'(x) \, dx \tag{7}$$

is obtained. Although this can be written alternatively as

$$dy = \frac{dy}{dx} \, dx,$$

it is obviously better to give (7) greater prominence.

The essential novelty of this treatment is that $dx$ does *not*, in the first instance, appear as the horizontal step $\overline{PN}$ in Fig. 11.1, but as the *vertical* step $\overline{N_0T_0}$ associated with the graph of $y = x$. The 'tangent' to this graph at $P_0$ is, of course, the line itself. In other words, the quotient $dy \div dx$ does not arise first as a quotient of vertical and horizontal steps associated with the curve $y = f(x)$ (i.e. *not* as $\overline{NT} \div \overline{PN}$), but as a *ratio of two vertical steps* ($\overline{NT} : \overline{N_0T_0}$), a ratio that compares a property of the given curve $y = f(x)$ with the corresponding property of the 'reference' curve $y = x$. Of course, since $\overline{N_0T_0} = \overline{P_0N_0} = \overline{PN}$, $\overline{NT} : \overline{N_0T_0}$ then turns out to be the same as $\overline{NT} \div \overline{PN}$, but the difference of emphasis in the definition is crucial to a sensible presentation of differentials.

### 11.5

In the final formula (7), the actual value of $\delta x$ chosen originally has become irrelevant. This is because the relation (7) is homogeneous in the differentials $dy$, $dx$, so that the *relation* between $dy$ and $dx$ is the same, whatever $\delta x$ is chosen. Nevertheless, *some* specific non-zero $\delta x$ must be nominated, in order to initiate the sequence of definitions.

Equation (7) is more general than (5), because (7) continues to hold if $x$ is not independent. If, say, $y = f(x)$ but $x = g(u)$ where $u$ is independent, then

$$dx = \frac{dx}{du} \, du$$

and, when $y$ is expressed as a function of the *independent* variable $u$,

$$dy = \frac{dy}{du} \, du.$$

Hence, by the chain rule,

$$dy = \frac{dy}{dx} \frac{dx}{du} \, du$$

and so

$$dy = \frac{dy}{dx} \, dx,$$

the relation being the same as it would have been had $x$ been independent.

It will perhaps be instructive to recapitulate the sequence in which the various quantities are defined.

(0) A function $f$ is given, together with a value of $x$ belonging to the domain of $f$.

(1) An increment $\delta x$ ($\neq 0$) is taken (with $x + \delta x$ also belonging to the domain of $f$).

(2) For the given $f$, the given $x$ and the given $\delta x$, $\delta y$ is defined by $\delta y = f(x + \delta x) - f(x)$.

(3) $\delta y/\delta x$ is obtained by division.

(4) The behaviour of $\delta y/\delta x$ as $\delta x \to 0$ is investigated.

(5) If, as $\delta x \to 0$, $\delta y/\delta x$ tends to a limit, $f$ is derivable at $x$ and $f'(x) = \lim_{\delta x \to 0} (\delta y/\delta x)$.

(6) Again, a specific increment $\delta x$ ($\neq 0$) is taken.

(7) For the given $f$, the given $x$ and the given $\delta x$, dy is defined by $dy = f'(x)\,\delta x$.

(8) With this definition, dy can be determined for *any* derivable function $f$ and, taking $f$ to be the identity function, $dx = \delta x$.

(9) Comparison of dy and dx leads to $dy = f'(x)\,dx$.

### 11.6

This process is not as excursive as it seems. Each step is well defined and quite logical. If an *exactly analogous* procedure is now applied to a function of more than one variable, if, for example, $w = f(x, y, z)$, one is led naturally to the relation

$$dw = f_1(x, y, z)\,dx + f_2(x, y, z)\,dy + f_3(x, y, z)\,dz,$$

or, if you prefer,

$$dw = \frac{\partial w}{\partial x}\,dx + \frac{\partial w}{\partial y}\,dy + \frac{\partial w}{\partial z}\,dz,$$

without any of the sleight-of-hand or mystery that so often surrounds such a statement.

In order to confine the geometry to 3 dimensions where visualization is easier, the function taken will be one of just 2 variables, but analytically the process extends quite straightforwardly to functions of more than 2 variables.

So $w = f(x, y)$, where $x$ and $y$ are independent variables, and $(x, y)$ is also used to represent a given number pair belonging to the domain of $f$. Independent increments $\delta x$ and $\delta y$ are taken in $x$ and $y$ respectively.

Fig. 11.2 shows a fragment of the surface $w = f(x, y)$, P being the point on the surface corresponding to the chosen number pair $(x, y)$ and Q that corresponding to $(x + \delta x, y + \delta y)$; the projections of P and Q on the $xy$-plane have been omitted for the sake of clarity. The rectangle PLNM with $\overline{PL} = \delta x$, $\overline{PM} = \delta y$ is horizontal (that is, parallel to the $xy$-plane) and $\overline{NQ}$ represents $\delta w = f(x + \delta x, y + \delta y) - f(x, y)$.

Suppose now that the surface is smooth enough for a tangent plane at P to exist. We shall investigate how that geometric assumption can be turned into an analytic condition on $f$.

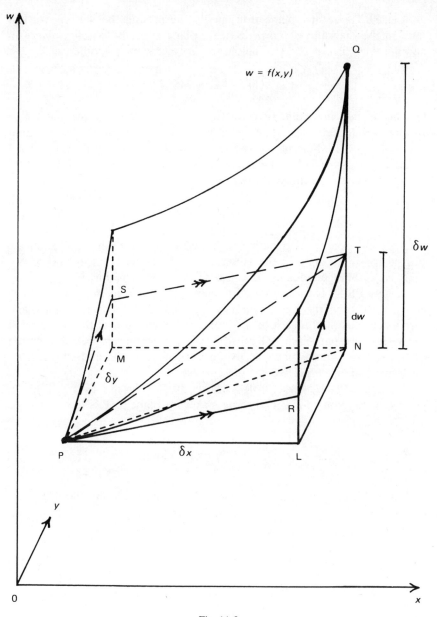

Fig 11.2

Let PRTS be any plane through P, intersecting the various vertical lines in the diagram as shown. Since a plane is cut by parallel planes in parallel lines, PRTS is a parallelogram. By elementary geometry,

$$\overline{NT} = \overline{LR} + \overline{MS}.$$

[If this is not immediately obvious to you, draw a construction line through R parallel to PM, meeting NT at U and observe that $\overline{NT} = \overline{NU} + \overline{UT}$ where $\overline{NU} = \overline{LR}$ and, since $\triangle RUT \equiv \triangle PMS$, $\overline{UT} = \overline{MS}$.] If the gradients of PR and PS within their own planes (referred to axes oriented like the given axes) are $l$ and $m$, then

$$\overline{NT} = l\,\delta x + m\,\delta y, \tag{8}$$

where $l$ and $m$ are independent of $\delta x$ and $\delta y$. [Alternatively, this can be proved by observing that the equation of any plane is linear and so $\overline{NT}$ must be a homogeneous linear combination of $\overline{PL}$ and $\overline{PM}$.]

If now PRTS is the supposed tangent plane at P to the given surface $w = f(x, y)$, then the line of intersection of PRTS with *any* other plane through P must be tangent to the curve of intersection of that plane with the surface. In particular, PT must be tangent to the arc PQ, the section of the surface by the plane PQN (and, of course, PR and PS must be tangents to their respective section curves).

Denote PN by $\delta r$ $(> 0)$, where $(\delta r)^2 = (\delta x)^2 + (\delta y)^2$. Then a necessary condition for PRTS to be a tangent plane at P is that

$$\frac{\overline{NQ}}{\overline{PN}} \to \frac{\overline{NT}}{\overline{PN}} \quad \text{as} \quad N \to P \quad \text{along} \quad PN,$$

and a sufficient condition is that this property should hold for *all* $\delta x$, $\delta y$. So

$$\frac{\overline{NQ}}{\delta r} = \frac{\overline{NT}}{\delta r} + \varepsilon \quad \text{where} \quad \varepsilon \to 0 \quad \text{as} \quad \delta r \searrow 0,$$

that is,

$$\overline{NQ} = \overline{NT} + \varepsilon\,\delta r$$

or, using (8),

$$\delta w = l\,\delta x + m\,\delta y + \varepsilon\,\delta r.$$

**11.7**

It is this necessary and sufficient condition for the existence of a tangent plane to the surface $w = f(x, y)$ at P that is used in the following analytic definition of the concept of the *differentiability* of the function $f$ at $(x, y)$.

*Definition* If $D \subseteq \mathbf{R} \times \mathbf{R}$ is an open set, a function $D \to \mathbf{R} : f$ is *differentiable* at $(x, y) \in D$ if, for all $\delta x$, $\delta y$ (such that $(x + \delta x, y + \delta y)$ also belongs to D),

$$f(x + \delta x, y + \delta y) - f(x, y) = l\, \delta x + m\, \delta y + \varepsilon\, \delta r, \tag{9}$$

where $(\delta r)^2 = (\delta x)^2 + (\delta y)^2$; $l, m$ are independent of $\delta x$, $\delta y$ (but depend, in general, on $x, y$); and $\varepsilon$ (which depends on $x, y, \delta x, \delta y$) tends to zero as $(\delta x, \delta y) \to (0, 0)$.

Perhaps, presented in this way, this very natural idea of the differentiability of a function of 2 variables will be more acceptable to you. Since the term $\varepsilon\, \delta r$ in (9) is a second order term, condition (9) is effectively saying that *a function is differentiable at a point if it is approximately linear there.*

In particular, taking $\delta y = 0$, $\delta x \neq 0$,

$$\left| \frac{f(x + \delta x, y) - f(x, y)}{\delta x} - l \right| = \left| \varepsilon\, \frac{\delta r}{\delta x} \right|$$

$$= |\varepsilon|,$$

and since, as $\delta x \to 0$, $\varepsilon \to 0$, it follows that

$$\frac{f(x + \delta x, y) - f(x, y)}{\delta x} \to l \quad \text{as} \quad \delta x \to 0.$$

By a similar argument,

$$\frac{f(x, y + \delta y) - f(x, y)}{\delta y} \to m \quad \text{as} \quad \delta y \to 0.$$

Hence,

*if $f$ is differentiable at $(x, y)$, then $f$ is derivable at $(x, y)$.*

As a corollary,

$$l = f_1(x, y), \qquad m = f_2(x, y).$$

This was, in fact, noted in the previous chapter: it was observed that, when they exist, $\partial w/\partial x$ and $\partial w/\partial y$ measure the gradients of PR and PS in their respective planes. Inserting these values of $l$ and $m$ into the definition (9), however, as some authors do, is not only unnecessary but actually obscures the underlying reason for introducing the concept.

To show that the converse implication is false, consider the function

$$\mathbf{R} \times \mathbf{R} \to \mathbf{R} : \phi : (x, y) \mapsto \sqrt{|xy|},$$

and its properties near $(0, 0)$.

Here, if $\delta x \neq 0$,

$$\frac{\phi(\delta x, 0) - \phi(0, 0)}{\delta x} = 0$$

and so

$$\phi_1(0, 0) = \lim_{\delta x \to 0} \frac{\phi(\delta x, 0) - \phi(0, 0)}{\delta x}$$

exists and is zero. Similarly $\phi_2(0, 0)$ exists, and hence $\phi$ is derivable at the origin.

Now suppose, if possible, that

$$\phi(\delta x, \delta y) - \phi(0, 0)$$

is expressible in the form

$$l\,\delta x + m\,\delta y + \varepsilon\,\delta r.$$

Then, taking $\delta y = 0$ (as in the above proof), $l = 0$ and, taking $\delta x = 0$, $m = 0$. So, since $l$ and $m$ have to be *inaependent* of $\delta x$ and $\delta y$,

$$l = m = 0.$$

But, writing $\delta x = \delta r \cos \theta$, $\delta y = \delta r \sin \theta$, $(\delta r > 0)$,

$$\phi(\delta x, \delta y) - \phi(0, 0) = \sqrt{|\delta x\,\delta y|}$$
$$= \delta r \sqrt{|\cos \theta \sin \theta|}$$

and hence

$$\varepsilon = \sqrt{|\cos \theta \sin \theta|}.$$

This is clearly inconsistent with the requirement that $\varepsilon \to 0$ as $\delta r \searrow 0$ and this contradiction proves that $\phi$ is *not* differentiable at the origin. The graph of the surface $w = \phi(x, y)$ is sufficiently interesting to be worth trying to visualize: when you have done so, the absence of a tangent plane at the origin will be apparent.

### 11.8

The extensions of these theories of derivability and differentiability to functions of more than 2 variables are entirely straightforward. But what information do they yield about a function of *one* variable?

Derivability at a point asserts the existence of a first derivative there: the definition of differentiability follows the pattern above, and says that the function is approximately linear.

*Definition* If $D \subseteq \mathbf{R}$ is an open set, a function $D \to \mathbf{R} : f$ is *differentiable* at $x \in D$ if, for all $\delta x$ (such that $x + \delta x \in D$),

$$f(x + \delta x) - f(x) = m\, \delta x + \varepsilon\, |\, \delta x\, |, \tag{10}$$

where $m$ is independent of $\delta x$ and $\varepsilon \to 0$ as $\delta x \to 0$.

This property implies, as before, that

$$\frac{f(x + \delta x) - f(x)}{\delta x} \to m \quad \text{as} \quad \delta x \to 0; \tag{11}$$

in other words, that $f$ is derivable at $x$ (and that $m = f'(x)$).

But, in *this* case, if $f$ is derivable at $x$, then the second equation (11) can be transformed into the first (10), so that $f$ is also differentiable at $x$.

So, whereas, in general, the two concepts of derivability (existence of all first order rates of change) and differentiability (approximate linearity; existence of a tangent space) are distinct, for functions of a single real variable the ideas are equivalent.

### 11.9

Let us return to the relation $w = f(x, y)$ and Fig. 11.2, and take up the challenge of extending the concept of a differential. The assumption that $f$ is a *differentiable* function will now be made. What is needed is an analogue of the step $\overline{\mathrm{NT}}$ in Fig. 11.1 and it is obvious that this will be provided by the step also called $\overline{\mathrm{NT}}$ in Fig. 11.2: it measures the step needed to get from N (at the same horizontal level as P) up to T on the tangent plane at **P**. It is this step $\overline{NT}$ that will be denoted by d$w$ and it is the formula for $\overline{NT}$ that will provide the analytic definition for d$w$. It has already been shown—see (8)—that

$$\overline{NT} = l\ \delta x + m\ \delta y,$$

where $l = f_1(x, y)$, $m = f_2(x, y)$. So, in accordance with this plan, the *definition*

$$dw = f_1(x, y)\ \delta x + f_2(x, y)\ \delta y \tag{12}$$

is now made, so that d$w$ shall be represented by $\overline{NT}$. Note that d$w$ is a function of 4 variables, $x, y, \delta x, \delta y$. If you prefer, you may think of (12) as saying

$$dw = \frac{\partial w}{\partial x}\ \delta x + \frac{\partial w}{\partial y}\ \delta y.$$

For example, if $w = x^4 \cos 3y$,

$$dw = d(x^4 \cos 3y) = 4x^3 \cos 3y\ \delta x - 3x^4 \sin 3y\ \delta y;$$

if $f(x, y) = (\ln x)/y, x = 2, y = -1, \delta x = 0{\cdot}3, \delta y = -0{\cdot}2,$

$$d\left(\frac{\ln x}{y}\right) = \frac{1}{2 \times (-1)} \times 0{\cdot}3 - \frac{\ln 2}{(-1)^2} \times (-0{\cdot}2) \simeq -0{\cdot}011$$

and so on. In particular, if $f(x, y) = x$,

$$d(x) = 1 \times \delta x + 0 \times \delta y,$$

that is,

$$dx = \delta x, \tag{13}$$

and, taking $f(x, y) = y$,

$$dy = \delta y. \tag{14}$$

Substituting (13) and (14) in (12),

$$dw = f_1(x, y)\ dx + f_2(x, y)\ dy, \tag{15}$$

or alternatively

$$dw = \frac{\partial w}{\partial x}\, dx + \frac{\partial w}{\partial y}\, dy.$$

Unlike total derivatives, partial derivatives are *not* quotients of differentials. But the linearity of relation (15) with respect to the differentials and its homogeneity should be particularly noted.

The planes $w = x$ and $w = y$ were not included in Fig. 11.2 in order to prevent too much clutter, but you might like to redraw Fig. 11.2 with these embellishments included so that it exactly resembles Fig. 11.1, showing all 3 *vertical* steps d$w$, d$x$, d$y$, as well as the horizontal increments $\delta x$, $\delta y$.

It would also be instructive to reconstruct, with the appropriate modifications for a function of 2 variables, the sequence of steps (0) to (9) in Section 11.5. Notice, however, that although step (5) requires $f$ to be merely derivable, step (7) requires $f$ to be differentiable and so the stronger hypothesis must be inserted before (7) and the consequent change in wording made in (8). The systematic nature of these definitions should not require any more emphasis and their further generalizations are immediate.

[It is just worth warning you, in passing, that in very old books you will sometimes see d$w$ described as a 'total differential'. This was done so that the quantities $(\partial w/\partial x)dx$ and $(\partial w/\partial y)dy$ could be called 'partial differentials'. But this was not a useful concept and these terms are no longer used: d$w$ is just a 'differential', as are d$x$ and d$y$.]

## 11.10

But just *why* are differentials held to be useful? It is partly because the formulae (7) and (15) and their extensions are linear and homogeneous in the differentials, but mainly because *they continue to hold even when the variables* $x, y, \ldots$ *are not independent*. This has already been discussed in reference to (7): now (15) will be examined.

In the remainder of this chapter, it will be assumed that all functions mentioned are differentiable at any point where their values are needed. Suppose first that $w = f(x, y)$, where $x = g(u, v)$, $y = h(u, v)$ and $u$ and $v$ are independent. Then

$$dx = \frac{\partial x}{\partial u}\, du + \frac{\partial x}{\partial v}\, dv \quad \text{and} \quad dy = \frac{\partial y}{\partial u}\, du + \frac{\partial y}{\partial v}\, dv. \qquad (16)$$

Elimination of $x$ and $y$ expresses $w$ as a function of the *independent* variables $u$ and $v$, so that

$$dw = \frac{\partial w}{\partial u} \, du + \frac{\partial w}{\partial v} \, dv.$$

The standard formulae for $\partial w/\partial u$ and $\partial w/\partial v$ give

$$dw = \left(\frac{\partial w}{\partial x} \frac{\partial x}{\partial u} + \frac{\partial w}{\partial y} \frac{\partial y}{\partial u}\right) du + \left(\frac{\partial w}{\partial x} \frac{\partial x}{\partial v} + \frac{\partial w}{\partial y} \frac{\partial y}{\partial v}\right) dv$$

and so, using (16),

$$dw = \frac{\partial w}{\partial x} \, dx + \frac{\partial w}{\partial y} \, dy.$$

Even though the variables $x$ and $y$ are not independent, the relation among the differentials $dw$, $dx$, $dy$ is the same as in the case when they were.

Again, if $w = f(x, y)$, where $x = g(t)$, $y = h(t)$ and $t$ is independent, then

$$dx = \frac{dx}{dt} \, dt \quad \text{and} \quad dy = \frac{dy}{dt} \, dt. \qquad (17)$$

Expressing $w$ as a function of the *independent* variable $t$ gives

$$dw = \frac{dw}{dt} \, dt,$$

and, using a well-known formula for $dw/dt$,

$$dw = \left(\frac{\partial w}{\partial x} \frac{dx}{dt} + \frac{\partial w}{\partial y} \frac{dy}{dt}\right) dt,$$

whence, using (17),

$$dw = \frac{\partial w}{\partial x} \, dx + \frac{\partial w}{\partial y} \, dy,$$

the relation again being the same as in the case when $x$ and $y$ were independent.

In a more general case, if, say, $w = f(x, y, z)$, where each of $x, y, z$ is a function of $r$ and $s$, where each of $r, s$ is a function of $t$, where $t$ is a function of $\xi$ and $\eta$, and $\xi$ and $\eta$ are independent, then you can see that, starting with the independent variables $\xi$ and $\eta$ and working back through the chain, you will come eventually to

$$dw = \frac{\partial w}{\partial x} dx + \frac{\partial w}{\partial y} dy + \frac{\partial w}{\partial z} dz.$$

A general proof using matrices can fairly easily be constructed.

## 11.11

The scientist is fond of using differentials (especially in subjects like thermodynamics) because he can make valid statements involving differentials without having to decide (or even know) in advance whether or not his variables are independent. But there is an additional practical advantage in these relations because of their linearity and homogeneity. This can be illustrated by considering one of those popular but rather tedious tasks given to students to change the set of variables associated with an operator (usually Laplace's operator $\nabla^2$).

Suppose $w = f(x, y)$, where $x = e^{3u} \sin 2v$, $y = e^{3u} \cos 2v$. The first step is to express $\partial w/\partial x$ and $\partial w/\partial y$ in terms of $\partial w/\partial u$ and $\partial w/\partial v$ and this problem will be solved in 4 different ways.

*Method* 1 (The worst method). Solve for $u$ and $v$ in terms of $x$ and $y$ *before* derivation. The example was chosen so that this step would not be too severe but, as you will appreciate, it could be very difficult or even impossible to obtain an explicit solution. Yet this is often the only method that occurs to students.

$$u = \tfrac{1}{6} \ln(x^2 + y^2), \qquad \tan 2v = \frac{x}{y}.$$

Thus

$$\frac{\partial u}{\partial x} = \frac{x}{3(x^2 + y^2)}, \qquad 2 \sec^2 2v \frac{\partial v}{\partial x} = \frac{1}{y} \quad \text{and so} \quad \frac{\partial v}{\partial x} = \frac{y}{2(x^2 + y^2)},$$

$$\frac{\partial u}{\partial y} = \frac{y}{3(x^2 + y^2)}, \qquad\qquad\qquad\qquad\qquad\qquad \frac{\partial v}{\partial y} = \frac{-x}{2(x^2 + y^2)}.$$

Hence

$$\frac{\partial w}{\partial x} = \frac{\partial w}{\partial u}\frac{\partial u}{\partial x} + \frac{\partial w}{\partial v}\frac{\partial v}{\partial x} = \frac{1}{6(x^2 + y^2)}\left(2x\frac{\partial w}{\partial u} + 3y\frac{\partial w}{\partial v}\right)$$

$$= \tfrac{1}{6}e^{-3u}\left(2\sin 2v\frac{\partial w}{\partial u} + 3\cos 2v\frac{\partial w}{\partial v}\right)$$

$$\frac{\partial w}{\partial y} = \tfrac{1}{6}e^{-3u}\left(2\cos 2v\frac{\partial w}{\partial u} - 3\sin 2v\frac{\partial w}{\partial v}\right).$$

*Method* 2 (Another complicated method). Solving for $u$ and $v$ is avoided, so this solution is a little better than Method 1. Derive both the relations

$$x = e^{3u}\sin 2v \quad \text{and} \quad y = e^{3u}\cos 2v$$

wo $x$, giving

$$1 = e^{3u}\left(3\sin 2v\frac{\partial u}{\partial x} + 2\cos 2v\frac{\partial v}{\partial x}\right)$$

$$0 = e^{3u}\left(3\cos 2v\frac{\partial u}{\partial x} - 2\sin 2v\frac{\partial v}{\partial x}\right).$$

Solve these *linear* equations for $\partial u/\partial x$ and $\partial v/\partial x$ and repeat the method to give $\partial u/\partial y$ and $\partial v/\partial y$. Then perform the same manipulations as in Method 1.

*Method* 3 (A better method). Solve for $\partial w/\partial x$ and $\partial w/\partial y$ in terms of $\partial w/\partial u$ and $\partial w/\partial v$ *after* derivation. This only involves the solution of *one* pair of (*linear*) equations.

$$\frac{\partial w}{\partial u} = \frac{\partial w}{\partial x}\frac{\partial x}{\partial u} + \frac{\partial w}{\partial y}\frac{\partial y}{\partial u} = 3e^{3u}\left(\sin 2v\frac{\partial w}{\partial x} + \cos 2v\frac{\partial w}{\partial y}\right)$$

$$\frac{\partial w}{\partial v} = 2e^{3u}\left(\cos 2v\frac{\partial w}{\partial x} - \sin 2v\frac{\partial w}{\partial y}\right).$$

Solve these for $\partial w/\partial x$ and $\partial w/\partial y$, obtaining

$$\frac{\partial w}{\partial x} = \tfrac{1}{6}e^{-3u}\left(2\sin 2v\frac{\partial w}{\partial u} + 3\cos 2v\frac{\partial w}{\partial v}\right)$$

$$\frac{\partial w}{\partial y} = \tfrac{1}{6}e^{-3u}\left(2\cos 2v\,\frac{\partial w}{\partial u} - 3\sin 2v\,\frac{\partial w}{\partial v}\right).$$

*Method* 4 (Use of differentials).

$$dx = \frac{\partial x}{\partial u}\,du + \frac{\partial x}{\partial v}\,dv = e^{3u}(3\sin 2v\,du + 2\cos 2v\,dv)$$

$$dy = \qquad\qquad e^{3u}(3\cos 2v\,du - 2\sin 2v\,dv).$$

The equations to be solved (for $du$ and $dv$) are again *linear*, and

$$du = \tfrac{1}{3}e^{-3u}(\sin 2v\,dx + \cos 2v\,dy)$$

$$dv = \tfrac{1}{2}e^{-3u}(\cos 2v\,dx - \sin 2v\,dy).$$

Substitute these in

$$\frac{\partial w}{\partial x}\,dx + \frac{\partial w}{\partial y}\,dy = dw = \frac{\partial w}{\partial u}\,du + \frac{\partial w}{\partial v}\,dv,$$

and equate the coefficients of $dx$ and $dy$, giving

$$\frac{\partial w}{\partial x} = \tfrac{1}{6}e^{-3u}\left(2\sin 2v\,\frac{\partial w}{\partial u} + 3\cos 2v\,\frac{\partial w}{\partial v}\right)$$

etc., as before.

There is obviously not much to choose between Methods 3 and 4 on the score of brevity. Differentials are never *essential* to solve problems of this type, but the scientist who is accustomed to using them may be anxious to persuade his mathematics colleagues to initiate his students into their secrets. He also finds them useful in problems like the following one.

If   $w = f(x, y),$ (18)

$$dw = \left(\frac{\partial w}{\partial x}\right)_y dx + \left(\frac{\partial w}{\partial y}\right)_x dy,$$ (19)

and, if   $y = g(x, z),$ (20)

$$dy = \left(\frac{\partial y}{\partial x}\right)_z dx + \left(\frac{\partial y}{\partial z}\right)_x dz.$$ (21)

Elimination of d$y$ between (19) and (21) leads to

$$dw = \left[ \left(\frac{\partial w}{\partial x}\right)_y + \left(\frac{\partial w}{\partial y}\right)_x \left(\frac{\partial y}{\partial x}\right)_z \right] dx + \left(\frac{\partial w}{\partial y}\right)_x \left(\frac{\partial y}{\partial z}\right)_x dz. \quad (22)$$

But elimination of $y$ between (18) and (20) expresses $w$ as a function of $x$ and $z$ and then

$$dw = \left(\frac{\partial w}{\partial x}\right)_z dx + \left(\frac{\partial w}{\partial z}\right)_x dz. \quad (23)$$

Comparison of the coefficients of d$x$ in (22) and (23) gives

$$\left(\frac{\partial w}{\partial x}\right)_z = \left(\frac{\partial w}{\partial x}\right)_y + \left(\frac{\partial w}{\partial y}\right)_x \left(\frac{\partial y}{\partial x}\right)_z.$$

For example, in thermodynamics,

$$\left(\frac{\partial E}{\partial T}\right)_p = \left(\frac{\partial E}{\partial T}\right)_V + \left(\frac{\partial E}{\partial V}\right)_T \left(\frac{\partial V}{\partial T}\right)_p.$$

Having marshalled in considerable detail some of the arguments for and against using differentials and discussed a lot of the background to the subject, which the textbooks tend to ignore, you should be in a position to decide what place you think the topic should hold in your school curriculum and you may, of course, come to the conclusion that it should not be there at all. But, if you do include it, you should at least be able to avoid the sort of sophistry that this topic seems to engender.

## 11.12

Before ending this chapter, there is a final but elementary observation to be made, prompted by the last example above. Remember that the usual notation for a partial derivative is always incomplete. When a total derivative d$s$/d$p$ occurs, one knows exactly what has been going on: there must have been a relation (either in fact or in concept) that expressed $s$ as a function of $p$ and this was derived wo $p$ to give d$s$/d$p$. When, however, $\partial q/\partial m$ occurs, the circumstances are not so clear; there must have been a relation that expressed $q$ as a function of $m$ and at least one other variable, but what those other variables were the symbol $\partial q/\partial m$ does not by itself reveal. This is why, in the example above, it was essential to write $(\partial E/\partial T)_p$ and $(\partial E/\partial T)_V$ to make the meaning clear. Students often make mistakes with partial derivatives by using $\partial w/\partial x$ on one line to mean, say, $(\partial w/\partial x)_y$ and then on another using $\partial w/\partial x$ to mean $(\partial w/\partial x)_z$, without noticing that they have changed its meaning. In potential theory, for example, care is needed with $\partial V/\partial \phi$: if $\phi$ is a cylindrical coordinate, $\partial V/\partial \phi$ will mean $(\partial V/\partial \phi)_{\rho,z}$; when the same $\phi$ is a spherical polar coordinate, it will mean $(\partial V/\partial \phi)_{r,\theta}$.

One is, of course, spared the intolerable burden of having to write every partial derivative in a form such as $(\partial q/\partial m)_{k,u}$ by the use of certain natural conventions, but it as well to make these conventions explicit in work with pupils. To refer back to the previous example that was solved by 4 methods,

$$\frac{\partial w}{\partial x}, \quad \frac{\partial w}{\partial y}, \quad \frac{\partial w}{\partial u}, \quad \frac{\partial w}{\partial v}$$

were used there to mean

$$\left(\frac{\partial w}{\partial x}\right)_y, \quad \left(\frac{\partial w}{\partial y}\right)_x, \quad \left(\frac{\partial w}{\partial u}\right)_v, \quad \left(\frac{\partial w}{\partial v}\right)_u,$$

the variables occurring either as the pair $(x, y)$ or as the pair $(u, v)$. Had it been possible that $(\partial w/\partial x)_v$ or $(\partial w/\partial v)_y$ might also have been involved in that piece of work, much more care would have had to be exercised. It is only because it is obvious that these quantities will not be relevant to the investigation that custom allows one to use the simpler, but less precise, notation. Similarly, if plane polar coordinates were present, $\partial w/\partial \theta$ would always be taken to mean $(\partial w/\partial \theta)_r$; if $(\partial w/\partial \theta)_x$ were likely to be needed as well, the convention could not be used.

# Chapter 12

# Stationary points

---

## 12.1

One meets so many students who have imperfectly understood the classification of stationary points that a few words of clarification may be appropriate. They may help you to prevent your pupils picking up similar erroneous ideas.

Given a twice-derivable function $f$ for which $f'(a) = 0$, time and time again students will tell you that

(1) if $f''(a) > 0$, then $f(x)$ has a minimum when $x = a$,
(2) if $f''(a) < 0$, then $f(x)$ has a maximum when $x = a$,
(3) if $f''(a) = 0$, then $f(x)$ has an inflexion† when $x = a$.

Unlike some of the more subtle errors noted elsewhere in this book, the counterexamples that give the lie to statement (3) could hardly be more accessible: $f(x) = x^4$ and $-x^4$ at $x = 0$, for example. Yet class after class of pupils is allowed to acquire this curious misconception. If $f''(a) = 0$, the nature of the stationary point at $a$ is undetermined; it may be either a minimum or a maximum or an inflexion: this test gives no guidance at all.

The first observation worth making is that most boys and girls are taught the second derivative test far too early. It is very much sounder practice to encourage them to discriminate by investigating the sign of the first derivative in the neighbourhood of the stationary point. Not only is that a more fundamental and generally applicable method (since it works even when the second derivative vanishes), but, once one gets away from the simplest sort of polynomial, it is usually an easier method too—think, for example, of the sort of rational or trigonometric expression (e.g. for potential energy) that one is likely to be trying to minimize in elementary applied mathematics.

---

†The spelling with an $x$ of the words connexion, deflexion, inflexion and reflexion is preferred by most etymologists.

The most radical way of looking at these students' errors, however, is to say that they are failing to distinguish between a necessary and a sufficient condition. When presenting this theory, their teachers have obviously missed a golden opportunity to emphasize this distinction. For, if $f'(a) = 0$, then

$f''(a) = 0$ is a necessary (but not a sufficient)
        condition for $f(x)$ to have an inflexion when $x = a$,
$f''(a) > 0$ is a sufficient  (but not a necessary)
        condition for $f(x)$ to have a minimum  when $x = a$,
$f''(a) \geq 0$ is a necessary (but not a sufficient)
        condition for $f(x)$ to have a minimum  when $x = a$,

and so on. An understanding of necessary and sufficient conditions, which is so important for advanced mathematics, only comes to students gradually. Yet there are many opportunities in the elementary curriculum for introducing these ideas.

## 12.2

The study of stationary points provides such a useful fund of good examples for practice in elementary derivation and of simple problem questions that the topic (admittedly important) tends to acquire prominence even beyond its merits. The penalty we pay for our exploitation of this rich seam is that pupils become conditioned into thinking that calculus is the only tool available for solving problems on stationary values.

Two examples of this come readily to mind. The first is the pupils' preference for finding maxima and minima of expressions of the type $\pm p \cos x \pm q \sin x$ by calculus, instead of by writing them in one of the forms $r \sin(x \pm \alpha)$, $r \cos(x \pm \alpha)$. The second is their slavish use of calculus methods with quadratic polynomials when 'completion of the square' will usually give more information more quickly (for example, in many problems on projectiles).

The work on quadratics can, in fact, be given a sharper focus if it is linked to ideas like the 'positive definite' property, which is a very important idea in later work anyway.

The fundamental identity for the quadratic

$$f(x) = ax^2 + bx + c \qquad (a \neq 0)$$

is

$$f(x) = \frac{1}{4a} [(2ax + b)^2 + (4ac - b^2)].$$

This tells us that $f(x)$ is stationary when $x = -b/(2a)$, with stationary value $(4ac - b^2)/(4a)$, which is a minimum if $a > 0$ and a maximum if $a < 0$. (Particular cases are, of course, discussed first!)

This same identity, however, also gives the more valuable information that

if $4ac > b^2$ and $a > 0$, then $f(x) > 0$ for *all* $x$, in which case $f(x)$ is called
*positive definite*;

if $4ac > b^2$ and $a < 0$, then $f(x) < 0$ for *all* $x$, in which case $f(x)$ is called
*negative definite*;

if $4ac < b^2$, then $f(x) > 0$ for some $x$ and $f(x) < 0$ for some $x$, and $f(x)$ is
called *indefinite*.

Naturally, these results will be illustrated by cartesian graphs of $y = f(x)$ as a matter of routine.

The remaining cases will probably not be given names at this stage:

if $4ac = b^2$ and $a > 0$, then $f(x) \geqslant 0$ for all $x$, and $f(x)$ is called
*non-negative definite*;

if $4ac = b^2$ and $a < 0$, then $f(x) \leqslant 0$ for all $x$, and $f(x)$ is called
*non-positive definite*.

Such quadratics are usually classified together as *semidefinite*.

### 12.3

The following simple problem on stationary values deserves to be better known. It is a useful example to keep in your mental file for challenging your pupils on suitable occasions.

A circular lake has a path running round its circumference. A man who can swim at speed $v$ and run at speed $kv$ ($k > 1$) wishes to get from one point on the path to the diametrally opposite point as quickly as possible. How does he do this?

If you are not already familiar with this interesting problem, you are strongly urged to solve it for yourself before reading the analysis that follows.

The first point to make, of course, is that there is obviously no advantage in doing a bit of swimming, then a bit of running, then another bit of swimming and so on. It clearly suffices to suppose that the man only swims once. So assume that he runs along an arc AP of the path subtending an angle $2\theta$ at the centre O of the lake and then swims along the chord PB to arrive at the point B diametrally opposite

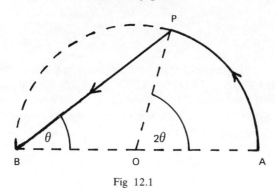

Fig 12.1

to A (Fig. 12.1). If $a$ is the radius of the lake, the time $t$ he takes is given by

$$t = \frac{2a\theta}{kv} + \frac{2a\cos\theta}{v}.$$

So

$$\frac{dt}{d\theta} = \frac{2a}{kv}(1 - k\sin\theta)$$

and

$$\frac{dt}{d\theta} = 0$$

when

$$\sin\theta = \frac{1}{k}.$$

At this point, many pupils announce that the man must run along an arc of the path subtending at the centre an angle $2\sin^{-1}(1/k)$ and swim along a chord subtending at the centre an angle supplementary to this and rule off happily.

Perhaps, however, *your* pupils will have been taught always to check the nature of the stationary points they have obtained, even when there is only one, and so, to humour you in this strange eccentricity, they will dutifully work out that,

$$\text{if} \quad \theta < \sin^{-1}\frac{1}{k}, \quad k\sin\theta < 1 \quad \text{and so} \quad \frac{dt}{d\theta} > 0;$$

$$\text{if} \quad \theta > \sin^{-1}\frac{1}{k}, \qquad k\sin\theta > 1 \quad \text{and so} \quad \frac{dt}{d\theta} < 0.$$

[For simplicity, it is being assumed that either they have not been introduced to the second derivative test or else they are mature enough to recognize for themselves that using it is usually laborious: in fact, in this case, that method is also quite straightforward, as you will see.]

Help! It is now clear to them that $\sin\theta = 1/k$ makes $t$ a maximum and *not* a minimum, so perhaps they are at least glad that they bothered to comply with that curious whim of yours. But their discovery does not solve the problem; it is obvious on common sense grounds that the man's journey time must have a lower bound, so what is it? It should not be too long now before they realize that this problem must be one in which the minimum time is attained not at a stationary point but at an end-point of a permitted interval and, noting the restriction $0 \leqslant \theta \leqslant \frac{1}{2}\pi$, and perhaps drawing a graph (Fig. 12.2) of $t$ against $\theta$, it will be clear to them that the man never gains any advantage by a combination of running and swimming.

If $\quad k < \frac{1}{2}\pi$, $\quad$ he must swim all the way, taking a time $\quad \dfrac{2a}{v}$ ;

if $\quad k > \frac{1}{2}\pi$, $\quad$ he must run all the way, taking a time $\quad \dfrac{\pi a}{kv}$ .

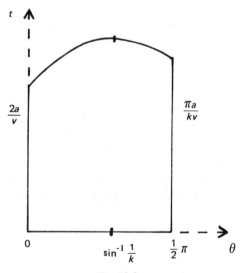

Fig 12.2

The merit of this problem is that the conditions imposed provide a natural restriction on the domain for $\theta$. It is, therefore, more satisfactory than those rather artificial problems that ask for the minimum value of, say, $3x^3 - 4x^2 - x - 3$ for $x \in [-1, 2]$, where one has to look at the value of the expression at the minimum turning point $(1, -5)$ and compare it with its values at the end-points $(-1, -9)$ and $(2, 3)$ in order to decide what the overall minimum is for the given closed interval.

It should be pointed out for completeness that the lake problem can in fact be solved without any explicit use of calculus at all, just by common sense, and this is obvious once one has realized what the solution is. But although anyone who thinks of this argument *before* carrying out the analysis above must be most warmly congratulated on his unusual insight, it spoils the interest of the problem for our present (pedagogic) purpose, which is why it has been suppressed so far; most people, in fact, tackle this problem by making an algebraic investigation on the lines above.

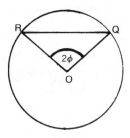

Fig. 12.3

If an arc QR subtends an angle $2\phi$ at the centre O of a circle, then

$$\frac{\text{chord QR}}{\text{arc QR}} = \frac{\sin \phi}{\phi},$$

which, for $0 < \phi < \frac{1}{2}\pi$ is a decreasing function of $\phi$. (This is being treated here as a standard result, but that hides the fact that calculus will almost certainly have been used in proving it.) So, if the swimming and running speeds are such that there is a time advantage in swimming along chord QR compared with running along arc QR, then this advantage will be even greater the larger $\phi$ becomes and the solution is seen to depend solely on the times the journey takes when completed using a single mode of locomotion.

**12.4**

No serious investigation of stationary values involving functions of 2 (or more) variables is likely to be undertaken at school, but, in spite of that, a misconception sometimes arises. The main classification of the stationary values of an expression $f(x, y)$ is a 3-way one: minima (*every* vertical plane section through the stationary point has a minimum there), maxima (similar) and saddle points (some sections through the point have minima there and some have maxima there). This trichotomy, as you will recall from your college work, is associated with the above-mentioned division of quadratics into positive definite, negative definite and indefinite.

Pupils, however, quite often pick up the impression that the saddle point is, in some sense, the 3-dimensional analogue of the inflexion. This attitude is completely misguided and should be vigorously discouraged. There *are* specialized types of stationary point in 3 dimensions that are outside the main 3-way classification, just as the inflexion is outside the dominant 2-way classification in the plane. For example, (1) if no part of a surface lies below a given plane but the surface touches that plane along a line, then no point of that line is a 'minimum' according to the definition; (2) a stationary point that has one or more plane sections containing an inflexion there is also a 'degenerate' stationary point. A comprehensive classification of such possibilities is laborious.

But the saddle point is in every sense a 'regular' or 'normal' stationary point, on exactly the same algebraic footing as the minimum and the maximum.

# Chapter 13

# Series expansions

---

## 13.1

The work done in sixth forms on (real) series expansions is one of the most tricky parts of the syllabus to explain well. It is not surprising that many teachers capitulate and do little more than (1) calculate the coefficients in an *assumed* expansion

$$f(x) = c_0 + c_1 x + \ldots + c_n x^n + \ldots$$

by the highly dangerous process of term-by-term derivation, obtaining

$$c_n = \frac{f^{(n)}(0)}{n!} \; ;$$

(2) use these formulae to write down an expansion for the given expression; and (3) either tell their pupils the range of values of $x$ for which this expansion is valid or, if the students have done some previous work on convergence, make the unwarranted assumption that the expansion will be valid whenever the series obtained converges.

In this way, pupils are introduced to the 8 or 9 standard series they are most likely to need, namely the series listed in Table 13.1, page 184, together with the routine transformations of the logarithmic series.

Those who intend to teach should realize how unsatisfactory this procedure is and what an unsound foundation it lays for pupils' future work in analysis.

## 13.2

The first step, that of term-by-term derivation, assumes that, if

$$f(x) = u_1(x) + u_2(x) + \ldots$$

## Table 13.1—Standard series

$$e^x = 1 + x + \frac{x^2}{2!} + \frac{x^3}{3!} + \ldots + \frac{x^n}{n!} + \ldots \qquad \text{for all } x \in \mathbf{R}$$

$$\operatorname{ch} x = 1 + \frac{x^2}{2!} + \frac{x^4}{4!} + \frac{x^6}{6!} + \ldots + \frac{x^{2n}}{(2n)!} + \ldots \qquad \text{for all } x \in \mathbf{R}$$

$$\operatorname{sh} x = x + \frac{x^3}{3!} + \frac{x^5}{5!} + \frac{x^7}{7!} + \ldots + \frac{x^{2n+1}}{(2n+1)!} + \ldots \qquad \text{for all } x \in \mathbf{R}$$

$$\cos x = 1 - \frac{x^2}{2!} + \frac{x^4}{4!} - \frac{x^6}{6!} + \ldots + (-1)^n \frac{x^{2n}}{(2n)!} + \ldots \qquad \text{for all } x \in \mathbf{R}$$

$$\sin x = x - \frac{x^3}{3!} + \frac{x^5}{5!} - \frac{x^7}{7!} + \ldots + (-1)^n \frac{x^{2n+1}}{(2n+1)!} + \ldots \qquad \text{for all } x \in \mathbf{R}$$

$$\ln(1 + x) = x - \frac{x^2}{2} + \frac{x^3}{3} - \frac{x^4}{4} + \ldots + (-1)^{n-1} \frac{x^n}{n} + \ldots \qquad \text{for } -1 < x \leq 1$$

$$\tan^{-1} x = x - \frac{x^3}{3} + \frac{x^5}{5} - \frac{x^7}{7} + \ldots + (-1)^n \frac{x^{2n+1}}{2n+1} + \ldots \qquad \text{for } -1 \leq x \leq 1$$

$$\sin^{-1} x = x + \frac{1}{2} \frac{x^3}{3} + \frac{1 \cdot 3}{2 \cdot 4} \frac{x^5}{5} + \frac{1 \cdot 3 \cdot 5}{2 \cdot 4 \cdot 6} \frac{x^7}{7} + \ldots$$

$$+ \binom{2n}{n} \frac{x^{2n+1}}{2^{2n}(2n+1)} + \ldots \qquad \text{for } -1 \leq x \leq 1$$

$$(1 + x)^a = 1 + ax + \frac{a(a-1)}{2!} x^2 + \frac{a(a-1)(a-2)}{3!} x^3 + \ldots$$

$$+ \frac{a(a-1) \ldots (a-n+1)}{n!} x^n + \ldots \qquad \text{for } -1 < x < 1^\dagger$$

for a certain set of values of $x$ and if $u_n(x)$ is derivable for all $n$ for a certain set of values of $x$, then, when $x$ belongs to the intersection of those sets,

$$f'(x) = u'_1(x) + u'_2(x) + \ldots ;$$

†For $-1 < x < 1$ for *all* $a$, but also for $x = 1$ if $a > -1$ and for $x = -1$ if $a > 0$. The binomial series is discussed further (with a different objective) in Section 9.5.

that is, that

$$\frac{d}{dx} \sum_{n=1}^{\infty} u_n(x) = \sum_{n=1}^{\infty} \frac{d}{dx} u_n(x).$$

Even so simple an example as

$$x - \frac{x^3}{3} + \frac{x^5}{5} - \frac{x^7}{7} + \ldots = \tan^{-1} x \qquad \text{for } -1 \leqslant x \leqslant 1,$$

$$1 - x^2 + x^4 - x^6 + \ldots = \frac{1}{1 + x^2} \qquad \text{for } -1 < x < 1 \text{ only,}$$

or the corresponding series for $\ln(1 + x)$ and $(1 + x)^{-1}$, will show you that care is needed. An example like

$$\sin x - \tfrac{1}{2} \sin 2x + \tfrac{1}{3} \sin 3x - \tfrac{1}{4} \sin 4x + \ldots = \tfrac{1}{2}x \qquad \text{for } -\pi < x < \pi,$$

where the series

$$\cos x - \cos 2x + \cos 3x - \cos 4x + \ldots,$$

obtained by term-by-term derivation wo $x$ is not even convergent (since $u_n(x) \nrightarrow 0$ as $n \to \infty$), will show you that there can be severe objections to this casual assumption.

It is the classic problem of having to perform two limit operations and having to decide whether the result will be the same irrespective of the order in which the operations are performed. At its most elementary level, it is the observation that

$$\lim_{a \to 0} \lim_{b \to 0} \frac{2a + 7b}{5a + 3b} = \frac{2}{5},$$

while

$$\lim_{b \to 0} \lim_{a \to 0} \frac{2a + 7b}{5a + 3b} = \frac{7}{3}.$$

If this example strikes you as too trivial or artificial to merit serious attention, try verifying the following results. Define

$$A_n(t) = 1 + \frac{1}{2^t} + \frac{1}{3^t} + \ldots + \frac{1}{n^t} - \int_1^n \frac{dx}{x^t} \qquad (t > 0).$$

Show that

$$\lim_{t \searrow 0} \left[ \lim_{n \to \infty} A_n(t) \right] = \frac{1}{2} \, ,$$

$$\lim_{n \to \infty} \left[ \lim_{t \searrow 0} A_n(t) \right] = 1 \, .$$

[Drawing the graph of $y = 1/x^t$ for $t > 0$ may give you some hints. Show that

$$\int_{r-1}^r \frac{dx}{x^t} < \frac{1}{2} \left[ \frac{1}{(r-1)^t} + \frac{1}{r^t} \right] \quad \text{and} \quad \int_{r-1/2}^{r+1/2} \frac{dx}{x^t} > \frac{1}{r^t} \, .$$

Deduce that

$$\frac{1}{2} \left( 1 + \frac{1}{n^t} \right) \ < \ A_n(t) \ < \ \frac{1}{1-t} \left[ 1 - \left( \frac{1}{2} \right)^{1-t} \right] + \frac{1}{2} \frac{1}{n^t} \, .$$

This inequality gives both results, although the second one is more simply obtained directly from the definition of $A_n(t)$.]

Similar questions arise in attempting to compare $\partial^2 w/\partial x \partial y$ and $\partial^2 w/\partial y \partial x$ and in changing the order of integration in an iterated integral, as well as in the term-by-term derivation and integration of series. Such problems are always fraught with particular difficulty.

### 13.3

The other assumption in the above treatment of series is that the interval of validity of the expansion, that is, the range of values of $x$ for which the sum of

$$f(0) + f'(0)x + \frac{f''(0)}{2!} x^2 + \dots$$

is $f(x)$ is the *same* as the interval of convergence. Fortunately—or, rather, unfortunately—this happens to be the case with the 9 series already noted and with certain other elementary series; it is unfortunate because it tempts pupils (and teachers) into making these rash assumptions.

The easiest way to make the distinction clear is to take the classic example of $e^{-1/x^2}$. For ease of printing, this will be written as $\exp(-1/x^2)$; there is no conflict here with the definition of $\exp z$ when $z$ is complex [see Section 8.5], since, if $z$ is $x$-axal, $\exp(x + 0i) = e^x$.

This example is one that all teachers should be familiar with in order to prevent them saying misleading things about Maclaurin expansions to their students. It is worthy of your careful study, as it indicates just what treacherous currents swirl beneath the apparently placid surface of the theory of these expansions. It is a pity that the mathematics is rather too difficult to pass on to pupils.

Let

$$\varepsilon(x) = \exp\left(-\frac{1}{x^2}\right) \qquad (x \neq 0).$$

As $x \to 0$, $\varepsilon(x) \to 0$ and so, by *defining*

$$\varepsilon(0) = 0,$$

a function $\mathbf{R} \to \mathbf{R} : \varepsilon$ is obtained that is continuous throughout its domain.

It is a standard property of the exponential function that, for all $r$,

$$\frac{1}{x^r} \exp\left(-\frac{1}{x^2}\right) \to 0 \quad \text{as} \quad x \to 0$$

and this property will be used freely in what follows.

Now, if $x \neq 0$,

$$\varepsilon'(x) = \frac{2}{x^3} \exp\left(-\frac{1}{x^2}\right),$$

and

$$\varepsilon'(0) = \lim_{h \to 0} \frac{\exp\left(-\frac{1}{h^2}\right) - 0}{h} = 0.$$

So

$$\varepsilon'(0) = \lim_{x \to 0} \varepsilon'(x)$$

and hence $\varepsilon'(x)$ is continuous at $x = 0$.

A straightforward induction argument will prove that, for $x \neq 0$,

$$\varepsilon^{(n)}(x) = P_n\left(\frac{1}{x}\right) \exp\left(-\frac{1}{x^2}\right),$$

where $P_n(t)$ is some polynomial in $t$ of degree $3n$. The induction hypothesis is that, with $x \neq 0$, this formula is correct for $n = k$. Then

$$\varepsilon^{(k+1)}(x) = P_k\left(\frac{1}{x}\right)\frac{2}{x^3}\exp\left(-\frac{1}{x^2}\right) - \frac{1}{x^2}P_k'\left(\frac{1}{x}\right)\exp\left(-\frac{1}{x^2}\right)$$

$$= P_{k+1}\left(\frac{1}{x}\right)\exp\left(-\frac{1}{x^2}\right),$$

where $P_{k+1}(t) = 2t^3\,P_k(t) - t^2\,P_k'(t)$ and is a polynomial in $t$ of degree $3(k + 1)$.

But the formula for $\varepsilon^{(n)}(x)$ is correct when $n = 1$. So, for all $n \geq 1$,

$$\varepsilon^{(n)}(x) = P_n\left(\frac{1}{x}\right)\exp\left(-\frac{1}{x^2}\right) \quad \text{for} \quad x \neq 0.$$

Another induction argument will now show that $\varepsilon^{(n)}(0) = 0$. This time, the induction hypothesis is that

$$\varepsilon^{(k)}(0) = 0.$$

Then

$$\varepsilon^{(k+1)}(0) = \lim_{h \to 0}\frac{\varepsilon^{(k)}(h) - \varepsilon^{(k)}(0)}{h} = \lim_{h \to 0}\left\{\frac{1}{h}P_k\left(\frac{1}{h}\right)\exp\left(-\frac{1}{h^2}\right)\right\} = 0,$$

using yet again the standard result mentioned earlier.

Since, as shown above, $\varepsilon'(0) = 0$, it follows by induction that, for all $n \geq 1$,

$$\varepsilon^{(n)}(0) = 0.$$

Thus *the values of $\varepsilon(x)$ and of all its derivatives are zero at $x = 0$.* Before being shown this striking result, most university students are prepared to conjecture that the *only* function with this property is the zero function. Putting it another way, if $f(x)$ has derivatives of all orders at $x = 0$, then the value of any derivative of $f(x) + \varepsilon(x)$ at $x = 0$ is exactly the same as that of the corresponding derivative of $f(x)$.

The Maclaurin expansion of $\varepsilon(x)$ is

$$0 + 0x + 0x^2 + \ldots .$$

This series is convergent for all $x$ with sum 0, but the sum will never be equal to $\varepsilon(x)$, except at $x = 0$. In other words, the interval of convergence of this Maclaurin series is **R**, but the interval of validity of the expansion is the single number, 0. Before pondering further the implications of this result, it will be interesting to look at the cartesian graph of $y = \varepsilon(x)$ in Fig. 13.1; it has an exceptionally flattened minimum at the origin and an asymptote $y = 1$.

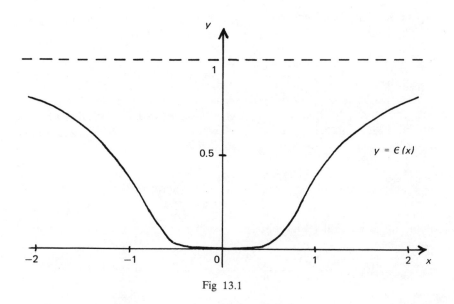

Fig 13.1

This dramatic discovery about the Maclaurin expansion of $\varepsilon(x)$ can be hammered home in a very vivid way. If $f$ is *any* function and $f(x)$ has a Maclaurin expansion

$$f(x) = c_0 + c_1 x + c_2 x^2 + \ldots ,$$

then $f(x) + \varepsilon(x)$ has *exactly the same* expansion

$$f(x) + \varepsilon(x) = c_0 + c_1 x + c_2 x^2 + \ldots .$$

The interval of convergence is the same for both (since they are the same series!), but there can be no point other than $x = 0$ where *both* expansions are valid.

**13.4**

The only logically acceptable way to deal with the problems raised by Taylor's and Maclaurin's† series, as you will remember from your analysis courses, is to start from Taylor's *theorem*, which expands a given expression in the form of a *finite* series *plus a remainder*,

$$f(a + x) = f(a) + f'(a)x + \ldots + \frac{f^{(n-1)}(a)}{(n-1)!} x^{n-1} + R_n,$$

and investigate the circumstances under which $R_n \to 0$ as $n \to \infty$.

The simplest form of remainder is Lagrange's form

$$R_n = \frac{f^{(n)}(a + \theta x)}{n!} x^n \quad \text{where} \quad 0 < \theta < 1,$$

but unfortunately this is not sharp enough even to deal with the binomial and logarithmic expansions when $-1 < x < 0$. Various other forms of remainder, such as Cauchy's form

$$R_n = \frac{(1 - \phi)^{n-1} f^{(n)}(a + \phi x)}{(n-1)!} x^n \quad \text{where} \quad 0 < \phi < 1,$$

or the more general (Schlömilch) form

$$R_n = \frac{(1 - \psi)^{n-p} f^{(n)}(a + \psi x)}{(n-1)! \, p} x^n \quad \text{where} \quad 0 < \psi < 1,$$

then have to be used, which adds considerably to the manipulative difficulties.

Indeed, even for the brightest students, it is hard to construct, without an altogether unreasonable expenditure of time, any treatment of series expansions that the pupils at school will find satisfying and convincing, that is not too difficult in a technical sense and that will not involve the teacher in making false statements that will have to be flatly contradicted later.

---

†Maclaurin's series is usually considered as a special case of Taylor's series, arising when $a = 0$, but it is equally possible to deduce Taylor's series from Maclaurin's because, if $F(x)$ is defined to be $f(a + x)$ and $F(x)$ has a Maclaurin series

$$F(x) = F(0) + F'(0)x + \ldots + \frac{F^{(n)}(0)}{n!} x^n + \ldots,$$

then, rewriting this in terms of $f$, Taylor's series emerges.

**13.5**

This leads one to wonder whether a study of series expansions in the above form is a worthwhile endeavour for sixth formers anyway and, once that not very revolutionary idea has surfaced, a possible solution immediately offers itself. What the student is really likely to need is a simple polynomial approximation to a given expression; he may want this for his applied mathematics, physics, economics, etc. What he is *unlikely* to require at this level is an infinite series expansion.

So one way to treat the subject may be to concentrate on obtaining the best-possible polynomial approximations of specified degree to a given expression in the region around $x = a$. Of course, one still has to agree what one is going to accept as 'best-possible', but that does not normally cause too much heart-searching. It is usually accepted quite readily that a reasonable criterion for this is that the polynomial and the given expression should share at $x = a$ the same values of as many derivatives as possible.

The tangent to a curve at a given point is the best-fitting line there, in the sense that it shares with the curve at that point the same value of $y$ and the same value of $y_1 [= dy/dx]$.

The circle of curvature at a given point of a curve is the circle that fits the curve there as closely as possible, in the sense that it shares with the curve at that point the same value of $y$, the same value of $y_1$ and the same value of $y_2$. [This, as you may already know, is the most natural way of obtaining the cartesian formula for the radius of curvature, $\rho$, at a point $(x, y)$ on a given curve. If the circle of curvature there has centre $(\alpha, \beta)$ and equation

$$(x - \alpha)^2 + (y - \beta)^2 = \rho^2,$$

then

$$x - \alpha + (y - \beta)y_1 = 0,$$

$$1 + y_1^2 + (y - \beta)y_2 = 0.$$

Here $y$, $y_1$, $y_2$ refer to the *circle*, but, if the circle is to share these quantities with the curve at $x$, then the values of $y$, $y_1$, $y_2$ in these equations will be those obtained from the given *curve*. Then

$$y - \beta = - \frac{1 + y_1^2}{y_2}, \qquad x - \alpha = \frac{y_1(1 + y_1^2)}{y_2},$$

(which gives, as a bonus, the coordinates $(\alpha, \beta)$ of the centre of curvature), and

$$\rho^2 = \frac{(1 + y_1^2)^3}{y_2^2},$$

giving $|\rho|$.]

Similarly, the polynomial of given degree $n$,

$$p(x) = c_0 + c_1 x + \ldots + c_n x^n,$$

that best approximates $f(x)$ in the neighbourhood of $x = a$ will be taken to be the one that has

$$p^{(k)}(a) = f^{(k)}(a) \quad \text{for} \quad k = 0, 1, \ldots, n,$$

(where, of course, $k = 0$ is to be interpreted as giving $p(a) = f(a)$).

This easily leads to the discovery that

$$c_k = \frac{f^{(k)}(a)}{k!} \quad \text{for} \quad k = 0, 1, \ldots, n.$$

To obtain these coefficients, only the derivation of a *polynomial* was necessary: there was no question of term-by-term derivation of a *series*.

Looked at in this way, these polynomials have a most striking property. If, having found the polynomial approximation of degree $n$, one then wishes to take instead the approximation of degree $(n + 1)$, the terms in $x^0$, $x^1$, ..., $x^n$ are exactly the same as before: all one has to do is to add an extra term $[f^{(n+1)}(a)]/[(n + 1)!]x^{n+1}$ without altering anything that has gone before. This really is a very dramatic and exciting result, yet it is rarely presented to pupils in this way!

University students will later meet a similar phenomenon with trigonometric approximations leading to Fourier series, which is worth recalling briefly here, even though the proofs will be omitted. A function $g$ is given with domain $(\alpha, \alpha + 2\pi)$. For simplicity, the reader can imagine that $g$ is continuous, although, in fact, $g$ can be allowed to have a finite number of bounded discontinuities in the interval. The problem is to find a trigonometric approximation to $g(x)$ of the type

$$q(x) = \tfrac{1}{2}a_0 + (a_1 \cos x + b_1 \sin x) + (a_2 \cos 2x + b_2 \sin 2x) + \ldots$$

$$+ (a_n \cos nx + b_n \sin nx)$$

of specified 'degree' $n$. For the present purpose, each pair of expressions within parentheses is considered as a single 'term'.

The aim in this problem (unlike the polynomial case) is not to make the approximation $q(x)$ fit $g(x)$ at any particular point of the domain, but rather to make it fit the given expression 'as well as possible' over the whole interval $(\alpha, \alpha + 2\pi)$. The appropriate criterion for this is of the usual 'least squares' type: in this case, that

$$\int_{\alpha}^{\alpha+2\pi} [g(x) - q(x)]^2 \, dx$$

shall be as small as possible. It can then be shown that, to achieve this minimum, it is necessary to take

$$a_k = \frac{1}{\pi} \int_{\alpha}^{\alpha+2\pi} g(x)\cos kx \, dx \qquad \text{for} \quad k = 0, 1, \ldots, n,$$

$$b_k = \frac{1}{\pi} \int_{\alpha}^{\alpha+2\pi} g(x)\sin kx \, dx \qquad \text{for} \quad k = 1, 2, \ldots, n.$$

When the coefficients are chosen in this way, the approximation $q(x)$ can be called the trigonometric approximation to $g(x)$ of degree $n$.

These approximations enjoy the same unexpected feature as the polynomials earlier: to obtain trigonometric approximations of successively higher degree, new terms are added one by one, but none of the earlier terms ever has to be altered. And, in just the same way as the polynomial approximations build up a Taylor series, so the trigonometric approximations build up a Fourier series.

## 13.6

If the teaching approach suggested were to be adopted, one would never actually be dealing with a Taylor *series* at this stage. One would just point out that the successive polynomial approximations to $f(a + x)$, namely

$$f(a) + f'(a)x$$

$$f(a) + f'(a)x + \frac{f''(a)}{2!} x^2$$

$$f(a) + f'(a)x + \frac{f''(a)}{2!} x^2 + \frac{f^{(3)}(a)}{3!} x^3\dagger$$

†Dashes are *not* a good notation for derived functions beyond $f''$, although $f'''$ is just about acceptable. But the incredibly hideous notation $f^{iv}$, $f^v$, ... to be found in nineteenth century textbooks should be buried out of sight! (It is not quite extinct.)

have the elegant property that a new term can be added to the best-possible approximation of degree $n$ to give the best-possible approximation of degree $(n + 1)$, without changing any of the terms in the original polynomial.

Incidentally, although the function $\varepsilon$ discussed above would not be used as a sixth form example, it is worth considering how our results about $\varepsilon$ fit in. They tell us that the polynomial that best represents $\varepsilon(x)$ in the neighbourhood of $x = 0$ is always the zero polynomial; no other polynomial, whatever its degree, can be a better approximation than this.

If this plan were followed, the step from writing the above polynomials to writing

$$f(a + x) = f(a) + f'(a)x + \ldots + \frac{f^{(n)}(a)}{n!} x^n + \ldots$$

is one that would not be attempted at this level. The crucial novelty of this last statement is the '$+ \ldots$' at the end. This is an embellishment that is all too lightly added by students—in many contexts—without always considering its tremendous significance. For the majority of pupils, this extension would be left to college or university courses, where Taylor's *theorem* (with remainder) can be properly studied and applied.

The above suggestion is controversial and many budding teachers may still wish to offer some presentation of series expansions, even to their pupils of average ability. Such praiseworthy endeavours are certainly not to be discouraged. Perhaps the above discussions, however, may help you to clarify your ideas about the problems latent in this topic and to make your own lessons mathematically sound.

### 13.7

Lest there be any lingering uncertainty about the writer's own feelings, let him make it quite clear that it is *not* the study of infinite series and convergence at the school level that he wishes to play down. Quite the opposite: he happens to believe very strongly that much valuable work in this area can and should be done long before formal definitions with epsilons are appropriate. It is not just the convergence of the geometric series $\Sigma x^n$ when $|x| < 1$ and its behaviour for other values of $x$ that are worth investigating. Important ideas will come from the realization that $\Sigma 1/\sqrt{n}$ and $\Sigma 1/n$ are divergent even though their

$n$th terms tend to zero;[†] that if $\Sigma |u_n|$ converges then $\Sigma u_n$ converges but not conversely; that rearrangement of the order of the terms in a convergent series *may* change the sum (or even make it cease to converge),[‡] and so on. All these and many other topics can be discussed informally, but with quite an acceptable standard of rigour by skilful teachers. No, it is not work on the intervals of *convergence* of power series that may lay bad foundations. What should be avoided are rash remarks about the intervals of *validity* of expansions and the encouragement of a cavalier attitude towards the very delicate questions about uniform convergence that are involved in the term-by-term derivation of series. These issues are really beyond the capability of sixth formers to discuss, even at an informal level.

---

[†]An even simpler series to illustrate this, which is not always noticed by teachers, is $\Sigma u_n$ with

$$u_n = \ln \frac{n + 1}{n} .$$

Here $u_n \to 0$ and $\Sigma_1^N u_n = \ln(N + 1) \to \infty$. Another is $\Sigma v_n$ with

$$v_n = \sqrt{(n + 1)} - \sqrt{n} = \frac{1}{\sqrt{(n + 1)} + \sqrt{n}} .$$

[‡]The standard examples to illustrate this are usually based on

$$1 - \frac{1}{2} + \frac{1}{3} - \frac{1}{4} + \ldots \qquad\qquad = s(= \ln 2)$$

$$1 + \frac{1}{3} - \frac{1}{2} + \frac{1}{5} + \frac{1}{7} - \frac{1}{4} + \frac{1}{9} + \ldots = \frac{3}{2} s$$

$$1 - \frac{1}{2} - \frac{1}{4} + \frac{1}{3} - \frac{1}{6} - \frac{1}{8} + \frac{1}{5} - \ldots = \frac{1}{2} s$$

$$1 - \frac{1}{\sqrt{2}} + \frac{1}{\sqrt{3}} - \frac{1}{\sqrt{4}} + \ldots \qquad\qquad \text{converges}$$

$$1 + \frac{1}{\sqrt{3}} - \frac{1}{\sqrt{2}} + \frac{1}{\sqrt{5}} + \frac{1}{\sqrt{7}} - \frac{1}{\sqrt{4}} + \frac{1}{\sqrt{9}} + \ldots \text{diverges} \ (s_n \to \infty)$$

$$1 - \frac{1}{\sqrt{2}} - \frac{1}{\sqrt{4}} + \frac{1}{\sqrt{3}} - \frac{1}{\sqrt{6}} - \frac{1}{\sqrt{8}} + \frac{1}{\sqrt{5}} - \ldots \text{diverges} \ (s_n \to -\infty).$$

# Chapter 14

# Non-commutative operations

## 14.1

With many quantities whose multiplication is non-commutative, for example, matrices and vectors, the definition of the product is clearly formulated and universally agreed, so that no reader is ever in any doubt as to precisely what is meant by the matrix AB or the vector **a ∧ b**.

But when the binary operation—even if it is called 'multiplication'—is really composition (of, for example, symmetry operations,[†] permutations or general mappings), there is no such harmony. In these cases the operation of composition corresponds to 'followed by', but the result '$\sigma$ followed by $\tau$' is sometimes written $\tau\sigma$ (or $\tau \circ \sigma$) and sometimes $\sigma\tau$ (or $\sigma \circ \tau$). The fact that, on such a fundamental issue, no general agreement has so far been reached among mathematicians will indicate to you that neither convention has overwhelming superiority. Nevertheless, the confusion it causes is a matter for the most profound regret and, as will be explained presently, the consequences of the original decision are far-reaching.

Analysts, who always represent their mappings $f$ by a functional *prefix* notation, $x \mapsto f(x)$, naturally denote '$f$ followed by $g$' by $gf$, since $(gf)(x)$ suggests $g[f(x)]$. As a wild oversimplification, it may be said that topologists, in contrast, tend to prefer to indicate their mappings $f$ by a *postfix* notation, $x \mapsto xf$ (or $x \mapsto x^f$), and, once they have elected to do this, '$f$ followed by $g$' is automatically represented by $fg$, since $x(fg)$ then becomes associated with $(xf)g$. Algebraists, however, are divided: some follow the analysts in writing mappings on the left, so that the

<hr/>

[†]Remember that composition is an operation (binary). But, in many of the examples in this chapter, the quantities being composed are themselves operations (singular). There is further consideration of this point in Section 1.14.

composite $\sigma\tau$ means '$\tau$ followed by $\sigma$'; others favour the contrary usage and, for them, $\sigma\tau$ represents '$\sigma$ followed by $\tau$'. The problem has repercussions both in linear algebra and in topics like group theory; it affects eigenvectors and simultaneous equations, as well as trans-formation geometry and permutations.

## 14.2

The advantage of $\sigma\tau$ being '$\sigma$ followed by $\tau$' (at least for a mathematician whose language is Indo-European) is that the operations follow one another in the natural order of writing. (An Arabic or Hebrew writer is, of course, allowed the liberty of disagreeing with this argument.) Also, although more trivially, when discussing why 'followed by' is associative, writing the operations in succession from left to right in the same direction as the conventional mapping arrows point *seems* to make the associativity of the operation even more self-evident than with the opposite convention.

$$x \overset{\rho}{\mapsto} x\rho \overset{\sigma\tau}{\mapsto} x\rho\sigma\tau$$

$$x \overset{\rho\sigma}{\mapsto} x\rho\sigma \overset{\tau}{\mapsto} x\rho\sigma\tau.$$

[A useful homely example to illustrate why 'followed by' is associative but not commutative is to use $\rho$ to denote the action of putting on one's shirt, $\sigma$ for putting on one's pullover and $\tau$ for putting on one's jacket. The difference between $\sigma\tau$ and $\tau\sigma$ is vividly illustrated and the equivalence of $\rho(\sigma\tau)$ and $(\rho\sigma)\tau$ clearly shown—or of $(\tau\sigma)\rho$ and $\tau(\sigma\rho)$ with the opposite convention.]

On the other hand, in analysis the functional prefix notation is much too well entrenched to consider altering it and therefore the case for using the same convention for all other mappings also is so powerful that for many mathematicians it overrides any other argument. Another strong point in its favour will be mentioned later in this chapter.

Note that both conventions are found in the notations for derivation. When $w = f(x, y, z)$ and a partial derivative is written $\partial w/\partial y$, the derivation operation is being denoted by a prefix, $\partial/\partial y$. When a derived function is written $f_3$ [or $f_z$, but see Section 10.3], the subscripts are a sort of postfix notation. This means incidentally that, in the first instance,

$$\frac{\partial^2 w}{\partial x\, \partial y} \quad \textit{must mean} \quad \frac{\partial}{\partial x}\left(\frac{\partial w}{\partial y}\right),$$

that is, the derivative obtained by deriving *first* wo *y* and *then* wo *x*, whereas

$$f_{12}\ [f_{xy}]\quad must\ \text{mean}\quad (f_1)_2\ [(f_x)_y],$$

that is, the derived function obtained by deriving *first* wo *x* and *then* wo *y*. Books that say otherwise are wrong: but, of course, because of the equality of such mixed derivatives for general, well-behaved functions, the mistake passes unnoticed.

### 14.3

In addition to the general problem about the order in which a composite is written, a subsidiary difficulty arises when information on binary operations is displayed in tabular form. One often sees a non-commutative operation table with bare headings like Fig. 14.1, in which it is uncertain whether the entry in the row with heading *s* and column with heading *t* will be the product that is being written *st* or the product *ts*; that is, whether the left factor is being 'plotted' down or across. When the binary operation happens to be composition and there is the further uncertainty as to whether the writer is using *st* to mean '*s* followed by *t*' or '*t* followed by *s*', a 4-fold problem of disentanglement arises. It is sound practice always to give labels to the headings of such tables for non-commutative operations, either as in Fig. 14.2 or as in Fig. 14.3, and, whenever appropriate, to make clear statements in close proximity that say

(1) how the product tabulated will be written ($\sigma\tau$ or $\tau\sigma$),

(2) whether the entry represents '$\sigma$ followed by $\tau$' or '$\tau$ followed by $\sigma$'.

Notice, by the way, that one of these tabular problems could be avoided if mathematicians elected to 'plot' the left and right factors like cartesian coordinates, with the table arranged as in Fig. 14.4, and the product *st* entered at the 'point' with 'coordinates' $(s, t)$.

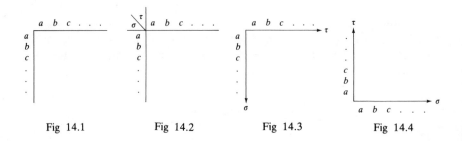

Fig 14.1          Fig 14.2          Fig 14.3          Fig 14.4

**14.4**

Just how far-reaching are the consequences of the original decision to write a mapping on the left or on the right may be judged from the few fragments of linear algebra quoted in Tables 14.1 and 14.2 (pages 200 to 203), comprising selected definitions and theorems enunciated according to each of the two conventions and written side by side for comparison.

Given vector spaces U, V, W over the same field F, the mappings from one space to another that are interesting are the *morphisms*,† that is, those that preserve the vector space structure. (In this context, morphisms are often called linear transformations, linear mappings or linear operators.) Such a mapping will be denoted here by a *script* capital letter, $\mathcal{T}$ or $\mathcal{S}$, as

$$U \rightarrow V : \mathcal{T} \quad \text{and} \quad V \rightarrow W : \mathcal{S}.$$

Scalars (elements of F) will be denoted by Greek letters, or, when elements of a matrix T, by $t_{ij}$; vectors by lower case Roman letters towards the end of the alphabet. Upper case Roman letters (except F, U, V, W) will be used for matrices, with X, $X^*$ denoting *column* matrices and Y, $Y^*$ denoting *row* matrices; T will denote the matrix representing the morphism $\mathcal{T}$ relative to some given pair of ordered bases in the spaces U, V.

On the left (Table 14.1) the morphism is written as a prefix $(x \overset{\mathcal{T}}{\mapsto} \mathcal{T}x)$; on the right (Table 14.2) as a postfix $(y \overset{\mathcal{T}}{\mapsto} y\mathcal{T})$. The suffixes $h, i, j, k$ have been uniformly assigned the following ranges:

| $h$ | $i$ | $j$ | $k$ |
|-----|-----|-----|-----|
| 1 to $l$ | 1 to $m$ | 1 to $n$ | 1 to $p$. |

A careful comparison of the subtle differences between the two developments will reveal various places where one scheme might be held to score over the other, but such advantages more or less cancel one another out. Notice that the definitions have been framed so that in each case the order of the matrix T in the first part of the tables will turn out to be $m \times n$. You will observe how this affects the way in which the $m$ and the $n$ have to be introduced into the theory originally $(F^n \rightarrow F^m : \mathcal{T}$ or $F^m \rightarrow F^n : \mathcal{T})$. If you have been brought up on one treatment and have to switch to the other, a very clear head is essential!

The other striking difference is that vectors represented on the left

†See Section 2.11.

### Table 14.1

*Mapping as prefix*

$$x \overset{\mathcal{T}}{\mapsto} \mathcal{T}x$$

(1)    If $U \to V : \mathcal{T}$ is a morphism, $\mathcal{T}(\alpha_1 x_1 + \alpha_2 x_2) = \alpha_1 \mathcal{T} x_1 + \alpha_2 \mathcal{T} x_2$, $(\alpha_1, \alpha_2 \in F)$.

(2)    If $U = F^n$ and $V = F^m$ are finite-dimensional and $(u_1, \ldots, u_n)$, $(v_1, \ldots, v_m)$ are ordered bases of $U$ and $V$ respectively, then, if

$$\mathcal{T}u_j = \sum_{i=1}^{m} t_{ij} v_i \quad (j = 1, \ldots, n) \quad (t_{ij} \in F),$$

the $m \times n$ matrix $T = [t_{ij}]$ is the matrix of $F^n \to F^m : \mathcal{T}$ relative to those bases.

(3)    If

$$x = \sum_{j=1}^{n} \xi_j u_j \quad \text{and} \quad \mathcal{T}x = \sum_{i=1}^{m} \xi_i^* v_i,$$

then

$$\xi_i^* = \sum_{j=1}^{n} t_{ij} \xi_j$$

or

$$X^* = TX,$$

where $X^*$, $X$ are the *column* matrices

$$X^* = \begin{bmatrix} \xi_1^* \\ \vdots \\ \xi_m^* \end{bmatrix}, \quad X = \begin{bmatrix} \xi_1 \\ \vdots \\ \xi_n \end{bmatrix}.$$

(4)    Let $V \to W : \mathcal{S}$ be another morphism, where $W = F^l$. Let the matrix of $\mathcal{S}$ relative to the *same* basis $(v_i)$ in $V$ and an ordered basis $(w_1, \ldots, w_l)$ in $W$ be the $l \times m$ matrix $S = [s_{hi}]$.

Then, relative to the basis $(u_j)$ in $U = F^n$ and $(w_h)$ in $W = F^l$, the composite transformation $U \to W : \mathcal{S}\mathcal{T}$ has the $l \times n$ matrix $ST$.

### Table 14.2

*Mapping as postfix*

$$y \overset{\mathscr{T}}{\mapsto} y\mathscr{T}$$

(1)    If $U \to V : \mathscr{T}$ is a morphism, $(\beta_1 y_1 + \beta_2 y_2)\mathscr{T} = \beta_1 y_1 \mathscr{T} + \beta_2 y_2 \mathscr{T}$, $(\beta_1, \beta_2 \in F)$.

(2)    If $U = F^m$ and $V = F^n$ are finite-dimensional and $(u_1, \ldots, u_m)$, $(v_1, \ldots, v_n)$ are ordered bases of $U$ and $V$ respectively, then, if

$$u_i\mathscr{T} = \sum_{j=1}^{n} t_{ij} v_j \quad (i = 1, \ldots, m) \quad (t_{ij} \in F),$$

the $m \times n$ matrix $T = [t_{ij}]$ is the matrix of $F^m \to F^n : \mathscr{T}$ relative to those bases.

(3)    If

$$y = \sum_{i=1}^{m} \eta_i u_i \quad \text{and} \quad y\mathscr{T} = \sum_{j=1}^{n} \eta_j^* v_j,$$

then

$$\eta_j^* = \sum_{i=1}^{m} t_{ij} \eta_i$$

or

$$Y^* = YT,$$

where $Y^*$, $Y$ are the *row* matrices

$$Y^* = [\eta_1^* \ldots \eta_n^*], \quad Y = [\eta_1 \ldots \eta_m].$$

(4)    Let $V \to W : \mathscr{S}$ be another morphism, where $W = F^p$. Let the matrix of $\mathscr{S}$ relative to the *same* basis $(v_j)$ in $V$ and an ordered basis $(w_1, \ldots, w_p)$ in $W$ be the $n \times p$ matrix $S = [s_{jk}]$.

Then, relative to the basis $(u_i)$ in $U = F^m$ and $(w_k)$ in $W = F^p$, the composite transformation $U \to W : \mathscr{T}\mathscr{S}$ has the $m \times p$ matrix $TS$.

### Table 14.1 (continued)

*Mapping as prefix*

(5)     If $V \rightarrow V : \mathscr{T}$ is an endomorphism of V, and $x \in V$ is a *non-zero* vector and $\lambda \in F$ a scalar, which together have the property that

$$\mathscr{T} x = \lambda x,$$

then $x$ is called an eigenvector of $\mathscr{T}$ and $\lambda$ an eigenvalue of $\mathscr{T}$.

(6)     If $V = F^n$ is finite-dimensional and $(v_1, \ldots, v_n)$ is an ordered basis of V relative to which $\mathscr{T}$ has the $n \times n$ matrix T and $x = \Sigma_j \xi_j v_j$, then

$$TX = \lambda X \quad \text{where} \quad X = \begin{bmatrix} \xi_1 \\ \vdots \\ \xi_n \end{bmatrix}.$$

The matrix X is a *column* eigenvector of T.

(7)     If T has $n$ linearly independent (column) eigenvectors $X_1, \ldots, X_n$ with corresponding eigenvalues $\lambda_1, \ldots, \lambda_n$, then the $n \times n$ matrix

$$P = [X_1 \, X_2 \ldots X_n]$$

has the property that the matrix $P^{-1}TP$ is diagonal:

$$P^{-1}TP = \begin{bmatrix} \lambda_1 & & & \\ & \lambda_2 & & \\ & & \ddots & \\ & & & \lambda_n \end{bmatrix}.$$

**Table 14.2 (continued)**

*Mapping as postfix*

(5)    If $V \to V : \mathcal{T}$ is an endomorphism of V, and $y \in V$ is a *non-zero* vector and $\lambda \in F$ a scalar, which together have the property that

$$y\mathcal{T} = \lambda y,$$

then $y$ is called an eigenvector of $\mathcal{T}$ and $\lambda$ an eigenvalue of $\mathcal{T}$.

(6)    If $V = F^n$ is finite-dimensional and $(v_1, \ldots, v_n)$ is an ordered basis of V relative to which $\mathcal{T}$ has the $n \times n$ matrix T and $y = \Sigma_j \eta_j v_j$, then

$$YT = \lambda Y \quad \text{where} \quad Y = [\eta_1 \ldots \eta_n].$$

The matrix Y is a *row* eigenvector of T.

(7)    If T has $n$ linearly independent (row) eigenvectors $Y_1, \ldots, Y_n$ with corresponding eigenvalues $\lambda_1, \ldots, \lambda_n$, then the $n \times n$ matrix

$$Q = \begin{bmatrix} Y_1 \\ Y_2 \\ \vdots \\ Y_n \end{bmatrix}$$

has the property that the matrix $QTQ^{-1}$ is diagonal:

$$QTQ^{-1} = \begin{bmatrix} \lambda_1 & & & \\ & \lambda_2 & & \\ & & \ddots & \\ & & & \lambda_n \end{bmatrix}.$$

by column matrices are represented on the right by row matrices and this affects the eigenvector theory in the second section of the tables, even though there the matrix T is square. For example, the matrix

$$T = \begin{bmatrix} 3 & 2 & 2 \\ 1 & -2 & 1 \\ -2 & 2 & -1 \end{bmatrix}$$

has eigenvalues

$$\lambda_1 = -3, \qquad \lambda_2 = 1, \qquad \lambda_3 = 2,$$

with corresponding column eigenvectors

$$X_1 = \begin{bmatrix} 0 \\ 1 \\ -1 \end{bmatrix}, \qquad X_2 = \begin{bmatrix} -1 \\ 0 \\ 1 \end{bmatrix}, \qquad X_3 = \begin{bmatrix} 10 \\ 1 \\ -6 \end{bmatrix},$$

or row eigenvectors

$$Y_1 = [1 \quad -4 \quad 1], \qquad Y_2 = [1 \quad 2 \quad 2], \qquad Y_3 = [1 \quad 1 \quad 1].$$

The matrices P and Q will then be

$$P = \begin{bmatrix} 0 & -1 & 10 \\ 1 & 0 & 1 \\ -1 & 1 & -6 \end{bmatrix}, \qquad Q = \begin{bmatrix} 1 & -4 & 1 \\ 1 & 2 & 2 \\ 1 & 1 & 1 \end{bmatrix},$$

with inverses

$$P^{-1} = \frac{1}{5}\begin{bmatrix} -1 & 4 & -1 \\ 5 & 10 & 10 \\ 1 & 1 & 1 \end{bmatrix}, \qquad Q^{-1} = \frac{1}{5}\begin{bmatrix} 0 & -5 & 10 \\ -1 & 0 & 1 \\ 1 & 5 & -6 \end{bmatrix},$$

giving

$$P^{-1}TP = \begin{bmatrix} -3 & 0 & 0 \\ 0 & 1 & 0 \\ 0 & 0 & 2 \end{bmatrix} = QTQ^{-1}.$$

The convention chosen also forces the selection of the row-echelon or the column-echelon form of a matrix as the primary target for its standard reduction.

## 14.5

When it comes to writing in matrix form a given set of simultaneous equations (or the equations of a given linear transformation), the work in the 2 cases has one striking practical difference.

*Mapping as prefix*

The equations are taken as

$$a_{11}x_1 + a_{12}x_2 + a_{13}x_3 = r_1 \Big\rbrace .$$
$$a_{21}x_1 + a_{22}x_2 + a_{23}x_3 = r_2 \Big\rbrace$$

In matrix form they become

$$\begin{bmatrix} a_{11} & a_{12} & a_{13} \\ a_{21} & a_{22} & a_{23} \end{bmatrix} \begin{bmatrix} x_1 \\ x_2 \\ x_3 \end{bmatrix} = \begin{bmatrix} r_1 \\ r_2 \end{bmatrix}.$$

*Mapping as postfix*

The equations are taken as

$$b_{11}y_1 + b_{21}y_2 + b_{31}y_3 = s_1 \Big\rbrace .$$
$$b_{12}y_1 + b_{22}y_2 + b_{32}y_3 = s_2 \Big\rbrace$$

In matrix form they become

$$\begin{bmatrix} y_1 & y_2 & y_3 \end{bmatrix} \begin{bmatrix} b_{11} & b_{12} \\ b_{21} & b_{22} \\ b_{31} & b_{32} \end{bmatrix} = \begin{bmatrix} s_1 & s_2 \end{bmatrix}.$$

The fact that the postfix convention requires the rectangular matrix occurring in the matrix equation to be the *transpose* of the array of coefficients that appears in the given set of equations is, for many teachers and lecturers, an overwhelming argument against the postfix convention. For those who are undecided, this observation, added to the fact that the prefix convention is universal in analysis, may well tip the balance in favour of writing *all mappings as prefixes*, in spite of the consequence that $\sigma\tau$ must then always stand for '$\tau$ followed by $\sigma$'.

## 14.6

To jump rather abruptly to another little conundrum involving non-commutative operations, this seems the appropriate place to mention an interesting type of fallacy, which your pupils may well 'invent'. It arises when they are attempting to construct the group table for a set of symmetry operations. It is not the sort of fallacy you deliberately show them in order first to mystify them and then to enlighten them: it is merely the sort of accidental mistake they can easily make themselves. Once made, however, it can be quite difficult for them to perceive why they are wrong and you may even find yourself momentarily stumped for an explanation unless you have met this type of paradox before.

An equilateral triangle PQR is given and the following symmetry operations are defined. An anticlockwise rotation of $\frac{2}{3}\pi$ about an axis through the centre of the triangle perpendicular to its plane is denoted by $\alpha$. Reflexions in the altitudes through P, Q, R are denoted by $p, q, r$ respectively. The effect of performing $\alpha$ followed by $p$ is to be investigated.

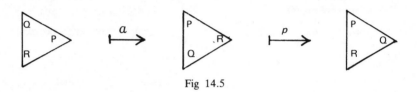

Fig 14.5

Labelling the triangle as shown in Fig. 14.5 and performing the operations indicated, the effect of $\alpha$ followed by $p$ appears to be the reflexion $r$. On the other hand, labelling the triangle as in Fig. 14.6, the result of $\alpha$ followed by $p$ now seems to be the reflexion $q$.

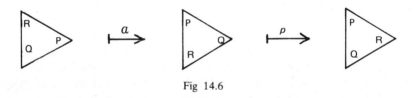

Fig 14.6

So the outcome seems to depend on the way the triangle is labelled!

You will admit that the above analysis does not look so very different from many perfectly respectable pieces of mathematics, which is why the learner can easily be led into error. You are, of course, invited to expose the fallacy for yourself clearly and thoroughly before reading on. Needless to say, it is not in any way dependent on the figure being a triangle; a similar paradox can arise in many problems.

The explanation is that the original operations are not well defined. The labels used to describe symmetry operations may *either* refer to a frame of reference attached to the triangle and moving with it *or* they may relate to a frame that stays fixed in the plane irrespective of the position of the triangle. The definitions offered were inadmissible, since they involved a mixture of the two frames.

If labels P, Q, R are attached to specific vertices of the triangle, then $\alpha$ must be defined according to how it is to affect P, Q, R: not by a description such as 'anticlockwise', which makes its effect dependent on the location of the triangle in the plane. If, on the other hand, $\alpha$ is related to the plane by a definition such as 'anticlockwise', then it is the points P, Q, R that must be defined with reference to $\alpha$, by saying whether PQR or RQP is to have the anticlockwise orientation. In this case, once specified, P, Q, R must remain fixed locations in the plane, unaltered by the transformations of the triangle.

Thus there are always two conventions that may underlie the construction of a group table of this type. Obviously, your pupils will only be introduced to one: it would be intolerable to confuse them by trying to explain both at this stage. But, as a teacher, you would be wise to acquaint yourself with both presentations. You yourself are likely only to have encountered one approach while a student and may be unaware that the other alternative exists. The group table for the symmetry operations of the equilateral triangle will, therefore, be constructed according to each convention. The results are shown on pages 208 to 211.

The important thing when you come to teach the topic is to decide which method you are going to follow (or work out which convention your textbook writer thinks he has adopted—this is not always as easy as it sounds, even when you have sorted out how he has plotted his table!), and then to explain your chosen method to your pupils.

### 14.7

To conclude this chapter, there is one technical point about the above problem (and ones like it) that will be appreciated by those of you who are knowledgeable about symmetry groups.

In constructing the symmetry group of the equilateral triangle, the operations $p, q, r$ were taken to be reflexions in the medians of the triangle, *not* half turns about those lines. This was strictly correct, for the following technical reason. When one is studying such a symmetry group as a 2-dimensional point group,† the operations to be considered must all be confined to the plane and must not involve lifting the triangle out of the plane into a space of higher dimension.‡ [Compare the remarks at the end of Section 7.6.] If you use a cardboard model to represent the mathematical triangle, then you must imagine that one face of the triangle is coloured, say, red and the other blue and that the

---

†A point group is a group of isometries all of which leave at least one point fixed.
‡Similarly, the cartesian graph of $x = f(y)$ is, *strictly* speaking, obtained from that of $y = f(x)$ by *reflexion* in the line $y = x$, *not* by rotation about that line, even though rotation provides an easier visual aid and would normally be used instead.

*Method* 1  Labels are attached to the triangle and move with it.

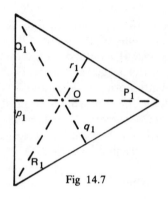

Fig 14.7

$P_1$, $Q_1$, $R_1$ denote specific vertices of the triangle and remain permanently attached to those vertices as the triangle is manipulated.

$p_1$, $q_1$, $r_1$ denote reflexions in the altitudes through $P_1$, $Q_1$, $R_1$, wherever those lines happen to be in the plane.

$\alpha_1$ denotes a rotation of the triangle through $\frac{2}{3}\pi$ about its centre O in the sense that takes the vertex $P_1$ to the position occupied by (say) the vertex $Q_1$, $Q_1$ to $R_1$ and $R_1$ to $P_1$.

[$\alpha_1$ may be either anticlockwise or clockwise according to the orientation of the triangle when the time comes to perform $\alpha_1$.]

$\alpha_1^3$ is the identity ($e$) and $\beta_1$ denotes $\alpha_1^2$ (which is the same as $\alpha_1^{-1}$).

With Method 1, for example,

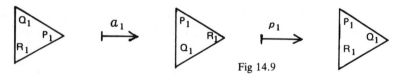

Fig 14.9

$\alpha_1$ followed by $p_1$ is equivalent to $r_1$.

The composite effect of each pair of operations is investigated in this

*Method* 2 Labels are fixed in the plane and are unaffected by the position of the triangle.

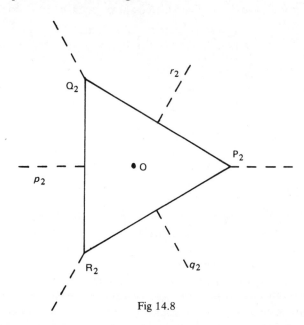

Fig 14.8

$\alpha_2$ denotes a rotation of the triangle about its centre O through $\frac{2}{3}\pi$ in a *definite* sense (anticlockwise, say) that is always the same.

$\alpha_2^3 = e$ and $\beta_2$ denotes $\alpha_2^2$ (or $\alpha_2^{-1}$).

$P_2$ denotes any one of the 3 points in the plane where a vertex of the triangle is situated. The image of $P_2$ under $\alpha_2$ is denoted by (say) $Q_2$ and the image of $Q_2$ by $R_2$.

$p_2$, $q_2$, $r_2$ denote reflexions in the lines $OP_2$, $OQ_2$, $OR_2$, which are fixed lines in the plane.

With Method 2, in order that progress can be followed, the vertices of the triangle *initially* positioned at $P_2$, $Q_2$, $R_2$ have been indicated in Fig. 14.10 below by X, Y, Z and here

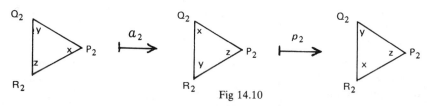

Fig 14.10

$\alpha_2$ followed by $p_2$ is equivalent to $q_2$.

The composite effect of each pair of operations is investigated in this

*Method 1 (continued)*

way, giving the table below, in which the entries describe the result of performing $\sigma$ *followed by* $\tau$. The suffixes have been omitted for clarity.

| $\sigma$ \ $\tau$ | $e$ | $\alpha$ | $\beta$ | $p$ | $q$ | $r$ |
|---|---|---|---|---|---|---|
| $e$ | $e$ | $\alpha$ | $\beta$ | $p$ | $q$ | $r$ |
| $\alpha$ | $\alpha$ | $\beta$ | $e$ | $r$ | $p$ | $q$ |
| $\beta$ | $\beta$ | $e$ | $\alpha$ | $q$ | $r$ | $p$ |
| $p$ | $p$ | $q$ | $r$ | $e$ | $\alpha$ | $\beta$ |
| $q$ | $q$ | $r$ | $p$ | $\beta$ | $e$ | $\alpha$ |
| $r$ | $r$ | $p$ | $q$ | $\alpha$ | $\beta$ | $e$ |

operations you have to investigate are the ones that do not interchange the colours. These are the 6 operations catalogued in the example, although 3 of them (the reflexions) are opposite isometries and so cannot be physically performed on the cardboard model.

With a young class, one is almost certain to replace $p$, $q$, $r$ defined above by half turns about the medians, so that the children can have the fun of performing the actual operations on a model, the model now being an *un*coloured triangle, with both faces assumed indistinguishable. The ambivalence exhibited in Methods 1 and 2 is, of course, still present when composing the operations: you *still* have to specify whether your operations are being defined relative to a moving frame or to a fixed frame. The group being studied in this case, however, is, technically, *not* the symmetry group of the (uncoloured) triangle, but its *rotation group*, which includes only the direct isometries. The full symmetry group of an object includes all opposite isometries as well, and in the case of this triangle is of order 12. Both of these are examples of 3-dimensional (*not* 2-dimensional) point groups.

This can be made clearer by setting out these 3-dimensional point groups in a table (using the familiar, but not entirely systematic, notation of Schoenflies).

|  | *Rotation group* | *Symmetry group* |
|---|---|---|
| Red–blue triangle | $C_3$ | $C_{3v}$ |
| Uncoloured triangle | $D_3$ | $D_{3h}$ |

*Method 2 (continued)*

way, giving the table below, in which the entries describe the result of
performing $\sigma$ *followed by* $\tau$. The suffixes have been omitted for clarity.

| $\sigma$ \ $\tau$ | $e$ | $\alpha$ | $\beta$ | $p$ | $q$ | $r$ |
|---|---|---|---|---|---|---|
| $e$ | $e$ | $\alpha$ | $\beta$ | $p$ | $q$ | $r$ |
| $\alpha$ | $\alpha$ | $\beta$ | $e$ | $q$ | $r$ | $p$ |
| $\beta$ | $\beta$ | $e$ | $\alpha$ | $r$ | $p$ | $q$ |
| $p$ | $p$ | $r$ | $q$ | $e$ | $\beta$ | $\alpha$ |
| $q$ | $q$ | $p$ | $r$ | $\alpha$ | $e$ | $\beta$ |
| $r$ | $r$ | $q$ | $p$ | $\beta$ | $\alpha$ | $e$ |

The groups $C_{3v}$ and $D_3$ are *isomorphic* groups of order 6 (with tables
as in the above example). But as symmetry groups they are reckoned
distinct, because the elements represent different physical operations.

The symmetry group of the triangle, which, as a 2-dimensional point
group, is usually called $D_3$, corresponds to the 3-dimensional group $C_{3v}$
and *not* to $D_3$. The notation is not as free from ambiguity as one could
wish, but this observation only serves to emphasize the fact that one
cannot talk about a symmetry group in absolute terms, but only in
relation to the space in which the operations are defined.

# Chapter 15

# Sets and set notations

---

## 15.1

'Doing sets' is the fashionable thing 'for children of all ages' (as they say on the toy boxes) and pupils may indeed find themselves studying aspects of sets at all ages from 6 to 16, and again at university. This may be admirable, but only provided there is a genuine progression in the lessons.

It will be as well to start by defining the scope of the work envisaged a little more closely. The objectives being pursued when teachers 'do sets' may be broadly divided into 4 categories:

(1) classification, (2) set language, (3) set algebra, (4) set theory.

(1) *Classification* is intended to describe the sort of activities typical of modern primary schools, using attribute blocks and hoops, giving the children perhaps some rudimentary ideas on intersection, union and complementation. This work has become a successful adjunct to that with numbers and shapes.

(2) The aim of studying *set language* is to encourage precision of expression. It is rightly held that in many branches of mathematics the vocabulary and symbolism of sets, set operations and set relations is the best vehicle for making accurate mathematical statements and it is claimed that systematic and careful adoption of this 'grammar' will encourage clear thinking and precise formulation of ideas.

(3) The term *set algebra* can best be used to describe the manipulative aspects of boolean† algebra as applied to sets. The value of

---

†The lower case initial letter in 'boolean' is not intended to show disrespect to that distinguished mathematician, George Boole (1810–1864). When an artificial adjective is constructed from a mathematician's name (as in euclidean space, newtonian mechanics, cartesian coordinate, pythagorean triplet, gaussian curvature, jacobian determinant, hermitean matrix, abelian group, etc), it is difficult to feel that the retention of the capital letter is still appropriate.

such a study is the broadening of pupils' horizons; making it clear that algebra is not just 'generalized arithmetic', as it was once called, and that there is nothing sacrosanct about the familiar axioms that apply to the operations on integers or to those on rational or real numbers. Set algebra is only one area where this cultural enrichment will be encouraged; geometric transformations, vectors and matrices will be other topics where the list of familiar integral domain or field properties will be found to need rewriting. It must be recognized, however, that set algebra can easily become an arid and purposeless activity, every bit as sterile as those notorious 'harder fractions', which are always pilloried nowadays as the scapegoat of earlier syllabuses. The difficulty is to do enough to bring out the potential of the algebra, but not so much that it becomes boring.

(4) *Set theory* is popularly used as a portmanteau term for any sort of study of sets. More correctly and usefully, however, it should be reserved for set theory as understood by universities and colleges: that is, the branch of mathematics pioneered by Georg Cantor (1845–1918), involving a study of the cardinality of infinite sets and the algebra of the cardinal numbers $\aleph_0$, $c$ and so on. It is not likely to be studied by many pupils at school, although the two basic results—(1) that the rational numbers are countable (denumerable) and (2) that the real numbers are not countable—are well within the capabilities of many sixth formers to prove. These demonstrations are valuable in their own right as stimulating examples of mathematical reasoning and the results give useful insight into important aspects of the rational and real numbers themselves.

All four of these types of investigation have mathematical value, although work in set algebra may cease to be challenging if carried too far. Work on sets is generally held to have been introduced into schools at the instigation of university mathematicians. But what those mathematicians were trying to encourage was a more widespread adoption of set *language*, not an interminable fiddling with Venn diagrams† at all ages from infancy to adolescence. All work on sets should be incidental to this main aim of encouraging precision of expression.

---

†There are other graphic representations for sets, such as Karnaugh maps, that have advantages over the ubiquitous Venn diagram, but these have not yet been much explored by teachers. Incidentally, the widespread habit of drawing Venn diagrams using exclusively *circles* is a matter for some slight regret. The boundaries need only be closed curves without self-intersections. They do not even have to be convex and, in fact, it is rather difficult (until you discover the trick) to make them all convex if 4 subsets have to be represented and impossible if there are more than four.

**15.2**

Unfortunately, set algebra is bedevilled by lack of agreement on notation and by the existence of many notations that are positively misleading for teaching purposes. One is painfully aware that many writers have only acquired a smattering of knowledge about boolean algebra before launching into print and have not thought very deeply about the notations they try to foist on the schools. How else can one explain someone advocating the use of '+' for the union operation and '−' for set difference? Few writers of school books, for example, seem to have studied the subject deeply enough even to have become acquainted with the associated boolean ring.

Two trivial points can be mentioned first. The symbol for the empty set is *not* the Greek letter phi ($\phi$), but a circle with an oblique line through it, $\varnothing$, a symbol that is reserved expressly for this purpose and should be distinguished, at least mentally, from $\phi$ (and also from that funny way, $\emptyset$, that computers are encouraged to type the digit zero). It was suggested some time ago—the writer unfortunately does not know by whom—that this symbol $\varnothing$ should be given the reading 'lero', a name that contained suggestions of the word 'zero' and of the German word 'leer', meaning 'empty'. This admirable proposal, however, never seemed to catch on. But it is well worth reviving.

Similarly, the symbol '$\in$' for the relation 'belongs to' should be distinguished from the Greek letter epsilon ($\varepsilon$ or $\epsilon$). Again, it is a special symbol devised (by Peano) to indicate the membership relation. Admittedly there is little to distinguish it in writing from the second form of epsilon above and inefficient printers sometimes substitute $\epsilon$ for $\in$, rather than having proper type cut; but, conceptually, they ought to be differentiated.

The first choice to be made is that of a label for the 'universe of discourse'. This is often denoted by E or $\mathscr{E}$ (from the French word 'ensemble' for set) and this is quite acceptable. What obviously *must* be avoided (but is not always) is the use of U for this 'universal set'. It is much too difficult to make a distinction in writing between U and the union sign $\cup$ and the work looks messy. There is, however (it seems to the writer), no objection to using I for this set. It plays after all the role of the unit element in the boolean algebra: for all $A \subseteq I$,

$$A \cap I = A, \qquad A \cup I = I.$$

The various notations for the singular operation[†] of complementation all have something to commend them. The best notation

†See Section 1.11.

for the complement of A is perhaps the commonest one, A$'$, but the alternative $\overline{\text{A}}$ is perfectly acceptable. Also popular, however, is the use of a functional prefix notation, $\complement$A or $\mathscr{C}$A, to denote the complement of A. This is not objectionable in itself, but the new symbol always seems to stand out obtrusively in this work, because the binary operations are symbolized using small infixes like $\cup$ in A $\cup$ B, rather than large prefixes like ‖AB. Also notice that, if $\complement$ is used, it is easy to print it in a different fount of type, but when written it resembles an ordinary capital letter C and hence looks momentarily as if it is going to denote some set, causing unnecessary jerks when reading.

## 15.3

The first real problem comes with the symbols for intersection and union.[†] The standard symbols ($\cap$ and $\cup$) for these operations, which are also due to Giuseppe Peano (1858–1932), were brilliantly chosen. Indeed, they are one of the happiest examples of symbolism in the whole of mathematics; they are beautifully simple and they clearly distinguish the two operations while vividly reminding us of the duality between them. This is further emphasized for English speakers by the well-chosen traditional readings 'cap' and 'cup' for the symbols. Occasionally, however, pupils have difficulty remembering which symbol is read as 'cap' and which as 'cup', although considering how many obvious reasons there are for the names, this is hard to understand. As mathematicians frequently point out, there might be advantages in adopting the alternative readings of 'and' for the intersection sign and 'or' for the union sign, which are, of course, suggested by their definitions:

$$x \in \text{A} \cap \text{B} \quad (\text{'A and B'}) \quad \text{means that} \quad x \in \text{A and } x \in \text{B},$$

$$x \in \text{A} \cup \text{B} \quad (\text{'A or B'}) \quad \text{means that} \quad x \in \text{A or } x \in \text{B}.$$

These would then correspond also to the readings of the related propositional connectives ($\wedge$ and $\vee$) in logic and emphasize important associations.

Because of the wish to keep the symmetry and not to give one operation precedence over the other, mathematicians commit themselves to a very heavy investment in brackets of various shapes. In a more familiar structure with two operations, a field (or ring) with addition and

[†]It is remarked in Chapter 1 (see Table 1.1) that the vocabulary for these set operations is not as complete as that for the arithmetic ones. The words 'intersection' and 'union' have to do duty both for the operations and for the results obtained when the operations are carried out.

multiplication, the operations are inherently lacking in symmetry—multiplication is distributive over addition: addition is not distributive over multiplication. By agreeing that one of the operations (multiplication) is to have precedence over the other (addition), one of the two expressions

$$(a + b) . c \quad \text{and} \quad a + (b . c)$$

(the second) can, by convention, be written *without* brackets. But, with

$$(A \cup B) \cap C \quad \text{and} \quad A \cup (B \cap C),$$

parentheses are required for both and, with more complicated expressions, intricate nests of brackets and braces are soon required, despite the associativity of both operations. It is undeniable that this gives such expressions a dismal appearance and makes manipulation a depressing occupation for everyone except the real enthusiast.

Because of this, several textbook writers have tried to take over the field symbols ($+$ and $\cdot$) for these set operations. There is, of course, illustrious precedent for this, as it was the notation Boole himself used. But whereas it may have been natural and reasonable for a gifted pioneer to take over familiar symbols and invest them with new and strange properties (rather than inventing completely new symbols), it is scarcely a practice that commends itself for the instruction of less mature intelligences. (A relic of this early history survives in books that call the union of two sets their 'logical sum' and the intersection their 'logical product'.)

What are the arguments in favour of $+$ and $\cdot$? There are two. Because of the traditional algebraic habit of giving multiplication precedence over addition, the set operation chosen for multiplication can be given similar precedence over the other operation and hence less brackets are needed. Also the lack of similarity in the visual appearance of the symbols '$+$' and '$\cdot$' (together with the reduction in the number of parentheses) makes, it can be claimed, the expressions *appear* less forbidding.

But the arguments against their adoption are numerous and overwhelming. First the duality between the symbols is lost and as a result the duality of the operations is mentally difficult to recapture, so strong are the emotional associations of the symbols '$+$' and '$\cdot$' with two operations that are *not* dually related in the way that '$\cup$' and '$\cap$' are.

There is no difficulty in deciding which of the operations $\cap$ or $\cup$ 'should' be denoted by multiplication. The analogies between the results

$$A \cap \varnothing = \varnothing \quad \text{and} \quad A \cap I = A$$

and

$$a \cdot 0 = 0 \quad \text{and} \quad a \cdot 1 = a$$

make intersection the obvious candidate. This means that union will be the operation symbolized by addition. This is in harmony with the analogy between

$$A \cup \varnothing = A$$

and

$$a + 0 = a,$$

but addition has nothing to offer that suggests the law

$$A \cup I = I,$$

which, written in the other way, would become

$$A + I = I.$$

This is the first place where the analogy breaks down. But, more seriously, only the first of the two dual distributive laws for sets

$$A \cap (B \cup C) = (A \cap B) \cup (A \cap C) \quad \text{and}$$

$$A \cup (B \cap C) = (A \cup B) \cap (A \cup C)$$

is recalled by

$$a \cdot (b + c) = a \cdot b + a \cdot c.$$

The other, in the $\{\cdot, +\}$ notation, takes the wildly unexpected form

$$A + B \cdot C = (A + B) \cdot (A + C).$$

The writer has met more than one very capable mathematician who learnt his set algebra from teachers who denoted union by addition and who has been absolutely incredulous when told that he could replace $(A + B) \cdot (A + C)$ by $A + B \cdot C$ or vice versa. His almost unshakable conviction has been that, *because* the symbol for addition has been used for union, *therefore* the union of sets must enjoy exactly the same properties as the addition of numbers.

There is also a third place where misunderstandings may arise. The use of both field operation symbols leads students to anticipate that their inverse operations will be possible within the system. This expectation is nurtured by the knowledge that complementation is an operation within the system and that, like the arithmetic operations of taking a negative and taking a reciprocal, this operation is involutory, that is,

$$A'' = A.$$

But, since a set and its complement must satisfy

$$A \cup A' = I \quad \text{and} \quad A \cap A' = \varnothing,$$

it is clear that a complement is an analogue of neither an additive inverse $(-a)$ nor a multiplicative inverse $(a^{-1})$, since

$$a + (-a) = 0 \quad \text{and} \quad a \cdot a^{-1} = 1.$$

There is a final and, to the writer, decisive reason for *not* selecting a notation that confuses union with addition and that is that there is another binary operation, not yet considered in this chapter, that is the *perfect* candidate for being called addition; this will be discussed presently. It is possible that this operation will not be given great prominence at school (although a case could be made out for making set algebra more interesting for pupils by increasing the variety of operations studied). But teachers (and textbook writers) should make themselves familiar with its properties so that they realize how misguided it is to symbolize union by addition, which is not only an inherently unsuitable notation, but thereby preempts a symbol that later can be more intelligently applied to a different binary set operation.

[There is actually one further argument from outside set algebra for not representing union as addition, although this one on its own is rather weak and it is also equivocal. In the theory of vector spaces, you will recall that, if A and B are both subspaces of a vector space V, then $A \cap B$ is also a subspace of V, but, in general, $A \cup B$ is not. (In fact, $A \cup B$ is not a subspace unless either $A \subseteq B$ or $B \subseteq A$.) But the set

$$A + B = \{a + b : a \in A, b \in B\}$$

*is* a subspace of V. Moreover it is entirely appropriate to use the overworked plus sign for *this* operation on subspaces. Students that are accustomed to write a union $A \cup B$ of sets as $A + B$ obviously have extra difficulty in appreciating what is going on.]

**15.4**

For those teachers, however, who still feel that the cap and cup notation with its numerous brackets is a deterrent to the enjoyment of set algebra, a compromise is possible, provided one is prepared to accept the loss of notational symmetry between the dual operations —and that is still a tremendously high price to pay.

It became clear during the above discussion that it was the addition rather than the multiplication sign that was responsible for the trouble; the intersection properties look more or less as 'expected' when $\cap$ is replaced by a multiplication sign. The idempotent property of intersection ($A \cap A = A$) certainly looks odd at first when written $A \cdot A = A$, but since this is just an *extra* property that this new sort of multiplication enjoys, its assimilation does not usually present any difficulty. It means that sets do not have 'powers', since, whatever the number of factors, $A \cdot A \cdot \ldots \cdot A = A$.

The compromise is to denote intersection by multiplication, *keep* the sign '∪' for union and give multiplication precedence over union. This reduces the number of brackets, yet minimizes the chance of confusion. The laws that looked peculiar when a plus sign was used for union no longer do so to quite the same extent when the notation $\{\cdot, \cup\}$ is employed. Thus pupils should be able to take quite cheerfully in their stride laws that say, for example,

$$A \cdot I = A \qquad\qquad A \cup \varnothing = A$$

$$A \cdot \varnothing = \varnothing \qquad\qquad A \cup I = I$$

$$A \cdot (B \cup C) = A \cdot B \ \cup \ A \cdot C \qquad A \ \cup \ B \cdot C = (A \cup B) \cdot (A \cup C)$$

$$A \cdot (A \cup B) = A \qquad\qquad A \ \cup \ A \cdot B = A$$

$$A \cdot A = A \qquad\qquad A \cup A = A$$

$$A \cdot A' = \varnothing \qquad\qquad A \cup A' = I$$

$$(A \cdot B)' = A' \cup B' \qquad\qquad (A \cup B)' = A' \cdot B'.$$

The loss of symmetry in the last two laws (de Morgan's rules) brings home perhaps even more forcefully than the loss in the other pairs of laws just how great is the price that has to be paid if one decides to use $\{\cdot, \cup\}$ instead of $\{\cap, \cup\}$. But at least that's better than using $\{\cdot, +\}$!

**15.5**

If other binary operations are to be studied, the next one will probably be *set difference*, denoted by \. This is defined by

$$A \setminus B = A \cap B',$$

that is,

$$x \in A \setminus B \quad \text{means that} \quad x \in A \quad \text{and} \quad x \notin B.$$

The best way of reading $A \setminus B$ is 'A but not B'.
  Since

$$I \setminus \varnothing = I \cap I = I, \qquad \varnothing \setminus I = \varnothing \cap \varnothing = \varnothing,$$

the operation $\setminus$ is not commutative. (Remember that one counter-example, however trivial, is sufficient to *dis*prove a conjecture.) Also, since

$$I \setminus (\varnothing \setminus I) = I \setminus \varnothing = I, \qquad (I \setminus \varnothing) \setminus I = I \setminus I = \varnothing,$$

the operation $\setminus$ is not associative.

Readers will probably have seen books in which set difference has been denoted by a minus sign $(-)$, instead of by the special sloping sign $(\setminus)$ designed for this operation. If used in an elementary book, this is another example of thoughtlessness or ignorance on the writer's part. In so far as the elements belonging to B are removed from A to obtain $A \setminus B$, the operation is reminiscent of subtracting the number $b$ from the number $a$ to give $a - b$, so that superficially the use of a minus sign might be held to be appropriate (and, indeed, it was used for set difference by the early pioneers). But there are snags and the snags are easily discovered. For one thing, although $a - (a - b) = b$, $A \setminus (A \setminus B)$ is clearly different from B (in general). More seriously,

$$A = B \ \Rightarrow \ A \setminus B = \varnothing, \quad \text{but} \quad A \setminus B = \varnothing \ \nRightarrow \ A = B.$$

So $A \setminus B = \varnothing$ is *not* equivalent to $A = B$ in the same way that $a - b = 0$ is equivalent to $a = b$, and therefore the use of a minus sign is highly misleading. This is precisely why a special sign $(\setminus)$ was devised, to avoid the prejudicial assumptions associated with the minus sign. For the same reason, the name 'set *difference*' for this operation was not an entirely happy choice, but it is well established and should only cause trouble to pupils who represent it by the same sign $(-)$ as that used for a difference between numbers. It is worth noting in conclusion that the binary relation between A and B *defined* by the condition $A \setminus B = \varnothing$ is, in fact, $A \subseteq B$; the possible binary relations between sets are investigated in Chapter 16.

**15.6**

The next operation usually considered is one whose usual name 'symmetric difference' is thoroughly badly chosen. It is often denoted by $\triangle$, which is a quite distinctive and satisfactory choice and will be used for the moment, although there is an even better notation, as you will discover presently. This operation is usually defined by

$$A \triangle B = (A \setminus B) \cup (B \setminus A) = (A \cap B') \cup (A' \cap B).$$

Using a distributive law and simplifying, the definition can easily be transformed into the equivalent forms

$$A \triangle B = (A \cup B) \cap (A' \cup B') = (A \cup B) \setminus (A \cap B),$$

using one of de Morgan's rules at the end. Hence

$$x \in A \triangle B \quad \text{means that} \quad x \in A \quad \text{or} \quad x \in B \quad \text{but} \quad x \notin A \cap B,$$

that is,

$$x \text{ belongs to A or to B but } not \text{ to both,}$$

so that, in distinction to $\cup$, which corresponds to the *inclusive* or, $\triangle$ corresponds to the *exclusive* or.

It is easily seen that $\triangle$ is commutative and, with a little more effort, it can be proved that $\triangle$ is associative.

$$A \triangle (B \triangle C) = [A \cap (B \triangle C)'] \cup [A' \cap (B \triangle C)]$$
$$= [A \cap \{(B \cap C) \cup (B' \cap C')\}] \cup [A' \cap \{(B \cap C') \cup (B' \cap C)\}]$$
$$= (A \cap B \cap C) \cup (A \cap B' \cap C') \cup (A' \cap B \cap C') \cup (A' \cap B' \cap C)$$

(using, on the second line, the alternative formula for $B \triangle C$ in $(B \triangle C)'$). The 3 sets A, B, C enter completely symmetrically into this expression, so that it is also equal to $C \triangle (A \triangle B)$ and hence, using the commutative law, to $(A \triangle B) \triangle C$. Further properties are

$$A \triangle \emptyset = A \qquad A \triangle I = A'$$

$$A \triangle A = \emptyset \qquad A \triangle A' = I.$$

Under neither $\cup$ nor $\cap$ does the set of subsets of I form a group. An identity element exists for each operation: this is $\varnothing$ for union $(A \cup \varnothing = A)$ and I for intersection $(A \cap I = A)$. But for neither operation do elements have inverses, since $A \cup B = \varnothing$ and $A \cap B = I$ are both impossible for a general set A. But under $\triangle$ the set of subsets of I *does* form a group (in fact, an abelian group), with identity element $\varnothing$ (since $A \triangle \varnothing = A$) and every element self-inverse (since $A \triangle A = \varnothing$).

Further, the operation of intersection is distributive over the operation $\triangle$. Note first that

$$(A \cap P) \setminus (A \cap Q) = (A \cap P) \cap (A \cap Q)'$$
$$= (A \cap P) \cap (A' \cup Q')$$
$$= (A \cap P \cap A') \cup (A \cap P \cap Q')$$
$$= \varnothing \cup (A \cap P \cap Q')$$
$$= A \cap (P \setminus Q),$$

so that $\cap$ is distributive over $\setminus$. Now

$$(A \cap B) \triangle (A \cap C) = [(A \cap B) \cup (A \cap C)] \setminus [(A \cap B) \cap (A \cap C)]$$
$$= [A \cap (B \cup C)] \setminus [A \cap (B \cap C)]$$
$$= A \cap [(B \cup C) \setminus (B \cap C)] \quad \text{by the previous result}$$
$$= A \cap (B \triangle C).$$

**15.7**

Now look at the structure of the set of subsets of I under the pair of operations $\triangle$, $\cap$. Under $\triangle$, the set forms an abelian group. The operation $\cap$ is associative and it is distributive over $\triangle$. The structure these two operations impose, therefore, is that of a *ring*.

It has already been noted that intersection can appropriately be represented as multiplication. It is now clearly seen that *the operation $\triangle$ is exactly analogous to an addition operation* and can (and will) be so represented henceforth.

$A \triangle B$    will be written    $A + B$    (read 'A plus B').

The use of the adjective 'symmetric' in 'symmetric difference' was never sufficient compensation to redeem the misleading use of the noun 'difference'. The best name for the operation is simply 'addition', with $A + B$ described as the 'sum' of the sets A and B. For most pupils, it seems unnecessary to introduce the inferior $\triangle$ symbol at all.

So the *intersection* of two sets may be described as their *product*, but a distinction *must* be preserved between the *union* of two sets (A ∪ B), corresponding to the inclusive or, and their *sum* (A + B), corresponding to the exclusive or. This now clearly demonstrates why the attempt by ill-informed writers to denote the union operation by a plus sign is such folly.

Since the multiplication operation (intersection) is commutative and since there is an identity element (I) for multiplication, the ring is in fact a *commutative ring with unit*. It is *not*, of course, a field, since, as already noted, there are no multiplicative inverses;[†] nor is it even an integral domain, since there are zero divisors: $A \cdot B = \varnothing$ does *not* imply that either $A = \varnothing$ or $B = \varnothing$.

Rewriting some of the properties mentioned earlier with the notation of addition will help to familiarize you, if you have been brought up on △, with the appearance of the rules with this symbolism. Note first that the set addition and multiplication operations are inherently *un*symmetric in their laws (unlike ∪ and ∩). Both operations are commutative and associative, but only multiplication is idempotent and there is just *one* distributive law, which has, however, a familiar look:

$$A \cdot (B + C) = A \cdot B + A \cdot C.$$

(Multiplication is, of course, being given precedence over addition—as it was over union also—to save brackets.) We have

$$A + \varnothing = A \qquad A + I = A'$$

$$A + A = \varnothing \qquad A + A' = I.$$

From these it follows that the sum, $A + A + \ldots + A$, of a number of equal sets A is $\varnothing$ if their number is even and A if their number is odd. Remember also that, since multiplication is idempotent, $A \cdot A \cdot \ldots \cdot A = A$, irrespective of the number of sets A in the product. Observe that neither $(A + B) \setminus B$ nor $(A \setminus B) + B$ is, in general, equal to A, which is yet another reason for *not* writing a set difference with an ordinary minus sign.

Because of the extra properties, such a ring is called a *boolean ring*. [Technically, this is a commutative ring with unit, which has the additional laws that, for all A, $A + A$ is the zero element (here $\varnothing$) and $A \cdot A = A$.]

---

[†] The boolean ring consisting of just 2 elements (0 and 1) *is* a field, but that is an exceptional case.

Compound sets become just polynomials. For example,

$$A \cup B = (A' \cap B')' = I + (I + A)(I + B) = I + I + A + B + AB$$
$$= A + B + AB.$$

Such polynomials are called *boolean polynomials*. They are manipulated just like ordinary polynomials, except that they have the extra properties noted. For example,

$$(I + B + AB)(A + B) = A + B + BA + BB + ABA + ABB$$
$$= A + AB$$
$$\text{(using } ABA = AAB = AB, AB + AB = \varnothing, \text{ etc)}.$$

Boolean rings are ideal structures for the careless; there are no such things as powers of an element and, since each element is self-inverse ($-A = A$), mistakes of sign are impossible!

## 15.8

The distinction mathematicians now make is between a *boolean algebra* (based on the operations of union, intersection and complementation) and a *boolean ring* (in which the operations emphasized are addition and multiplication). The distinction is convenient, but more apparent than real. Given a boolean algebra, the ring can be constructed as above by defining

$$A + B = (A \cup B) \cap (A' \cup B')$$
$$A \cdot B = A \cap B.$$

But conversely, given a boolean ring, with 1 as its unit element, the algebra can be derived from it by defining

$$A \cup B = A + B + AB$$
$$A \cap B = AB$$
$$A' = 1 + A.$$

Although the boolean structures have occurred here in connexion with set algebra, the above remarks really refer to the abstract boolean algebra and boolean ring and hence can be applied with equal force to the other two familiar examples of boolean structures: switching algebra and sentence logic (the propositional calculus). The links between all these investigations are the words 'and' and 'or'; indeed, boolean algebra may be characterized informally as the algebra of 'and' and 'or'.

George Boole himself was much exercised as to whether to take 'or' in its inclusive sense (Latin 'vel')[†] or its exclusive sense (Latin 'aut').[‡] He may have felt instinctively that the 'or' operation would have interesting properties whichever interpretation he took. One can express his dilemma epigrammatically by saying that Boole himself could not decide whether he wanted to invent boolean algebras or boolean rings. In practice, he tended to avoid the problem with 'or' by keeping his sets (or classes, as he called them) disjoint, so that the two interpretations ($x \cup y$ and $x + y$ above) would coincide.

It must be admitted that much of Boole's original work is far from lucid, although saying so is not intended to detract from his achievement. He tended to blaze away with his symbols, discovering that he could make interesting deductions using them, but rather conveniently ignoring expressions that had contradictory or indecisive interpretations, without being able to say clearly why some were being accepted and others rejected. Present-day treatments of boolean algebra owe much to the clarification provided by W Stanley Jevons (1835–1882).

**15.9**

Apart from the 4 binary operations already discussed ($\cap$, $\cup$, $\setminus$, $+$), there are 4 other essentially different binary operations between sets, but these are rarely thought worthy of mention in elementary work. The writer's preferred notations for these would be

$$A \uparrow B = \quad A' \cup B' \quad = (A \cap B)'$$

$$A \downarrow B = \quad A' \cap B' \quad = (A \cup B)'$$

$$A \to B = \quad A' \cup B \quad = (A \setminus B)'$$

$$A \leftrightarrow B = (A \to B) \cap (B \to A) = (A + B)',$$

[†]It is from the initial v in 'vel' that the sign v for the inclusive 'or' in logic is derived.
[‡]'Now whenever an expression involving these particles ["and" and "or"] presents itself in a primary proposition, it becomes very important to know whether the groups or classes separated in thought by them are intended to be quite distinct from each other and mutually exclusive, or not. Does the expression, "Scholars and men of the world", include or exclude those who are both? Does the expression, "Either productive of pleasure or preventive of pain", include or exclude things which possess both these qualities? I apprehend that in strictness of meaning the conjunctions "and", "or", do possess the power of separation or exclusion here referred to.... But it must at the same time be admitted, that the "jus et norma loquendi" seems rather to favour an opposite interpretation.' G Boole, *An investigation of the laws of thought*, page 55.

but numerous other notations will be found. These operations are examined in the next chapter. The 6 commutative operations ($\cap \cup + \leftrightarrow \uparrow \downarrow$) have been given signs with a vertical axis of symmetry; the non-commutative ones are $\setminus$ and $\rightarrow$.

# Chapter 16

# Boolean structure

---

## 16.1

Before investigating boolean operations and relations any further, it will be convenient to generalize the presentation. Although some of the observations in this chapter will range beyond the topics likely to be attempted at school, it is hoped that the discussion will provide a useful frame of reference for you, the teacher, because some important fundamental questions are going to be examined. The remarks here are a necessary preliminary to those on implication in Chapter 17.

The preoccupation with set algebra in the previous chapter gave rather too much emphasis to this particular concrete system at the expense of the abstract boolean structure, and anyway it will be useful to have the freedom to illustrate the properties of boolean algebras and rings in *all three* of the practical systems for which they are the mathematical model: (1) switching algebra, (2) sentence logic, (3) set algebra.

The zero and unit elements can then be denoted by 0 and 1 (rather than $\varnothing$ and I), which it was not desirable to do when dealing exclusively with sets. This is the usual notation in switching algebra also, where 0 normally denotes an open switch and 1 a closed switch. In applications to sentence logic, 0 will be associated with F (false) and 1 with T (true).[†] The notation for elements of the general algebra will be $x, y, z$, rather than the upper case letters usual for sets and used in Chapter 15. In switching algebra, the letters used for circuit variables will be $a, b, c$; in sentence logic, the letters for propositional (sentence) variables will be the traditional $p, q, r$.

---

[†]For certain specialized purposes in logic, it is sometimes more convenient to associate 0 with T and 1 with F. (This has the contingent effect that, in constructing the ring, multiplication has to be associated with ∨ rather than ∧, and addition with ↔.) But this option will be ignored in what follows, as being irrelevant for school teaching.

The singulary operation of the algebra, called complementation for sets, is usually also called complementation for switches, but is called negation in logic. The notation for the complement of switch $a$ will be $a'$; (the alternative $\bar{a}$ is just as convenient and is very commonly used): the switch $a'$ is closed when $a$ is open and vice versa. In logic, the negation† of $p$ is variously written

$$\rightharpoondown p, \rceil p, \sim p, \tilde{p}, \bar{p}, p', \mathsf{N}p;$$

the dash is not as common as some of the other notations, but will be used in this book for greater uniformity: the sentence $p'$ is true when $p$ is false and vice versa.

Obviously, any binary operation can be applied to pairs of elements of each system to create new elements with interesting properties. Traditionally, however, certain binary operations have been exploited more in some systems than in others. For example, the conditionality operation $\rightarrow$ is much more prominent in logic than in either switching or set algebra.

Four operations that are important in all the concrete systems are compared in the following table.

| | Switching algebra | Sentence logic | Set algebra |
|---|---|---|---|
| $\wedge\ (\cdot)$ | series connexion of switches | conjunction ('and') | intersection |
| $\vee$ | parallel connexion of switches | disjunction (inclusive 'or') | union |
| $+$ | 2-way (staircase) switch | addition (exclusive 'or') | addition |
| $\rightarrowtail\ (\backslash)$ | 'call-hold' operation‡ | 'but not' | set difference |

The one noteworthy distinction between set algebra and the other two systems is that a variable in both switching algebra and sentence logic is necessarily 2-valued—it can only take the values 0 and 1 (open or closed; false or true)—whereas in set algebra a letter can stand for any subset of the postulated universal set I. Actually, the difference is less than it appears to be at first sight: it rather depends on what one

| $x + y$ | $x$\$y$ | 0 | 1 |
|---|---|---|---|
| | 0 | 0 | 1 |
| | 1 | 1 | 0 |

| | $x$\$y$ | 0 | 1 | |
|---|---|---|---|---|
| | 0 | 0 | 1 | $x \vee y$ |
| | 1 | 1 | 1 | |

means by the word 'variable'. Some reconciliation will be effected later;§ for the present, the distinction will be allowed to stand. In the 2-valued

†Never call this 'the negative'.
‡$a \rightarrowtail b$ operates when $a$ is activated and is cancelled when $b$ is activated.
§See Section 17.19.

systems, notice particularly that the operation of addition corresponds *exactly* to the arithmetic operation of 'addition modulo 2'—unlike the operation v, which in *no* way resembles addition—providing yet more evidence (if more be needed) to discomfit those who try to confuse union or disjunction with addition.

## 16.2

The notations used in this book for the binary operations are listed in column 1 of Table 16.1, which gives a concordance showing the plethora of competing symbols you are likely to come across. Some of them (columns 2 and 3) are more or less acceptable substitutes for the ones being used here, but those in columns 4 and 5 are demonstrably bad (certainly in any elementary presentation of the subject) and should be zealously avoided by teachers. The notations in column 2 are ones often used in set algebra and in logic and are entirely satisfactory alternatives. Those in column 3 will be found in various books, but they do not seem to offer any advantages over the recommended symbols and are better avoided unless you feel you can argue strongly in favour of any of them. But there are powerful objections to all the infixes in columns 4 and 5. The arguments against using + (for 'or') and − (for 'but not') are presented at length in Chapter 15, while × is the standard notation for the cartesian multiplication of two sets and is thoroughly unsuitable for 'iff'. The authors of one book have perversely used ∧ for the exclusive or, but this must be regarded as an eccentricity not to be imitated, as must another gentleman's extraordinary use of ÷ for this operation. The symbol '⊃' for 'only if', however, is sanctioned by the authority of *Principia mathematica*, no less, and many writers since have followed Whitehead and Russell. But ⊃ is a notation greatly inferior to → for educational purposes, because the symbol '⊃' is in standard use for the *relation* of 'strictly contains' between sets. Since set algebra and sentence logic are both subject to the same (boolean) algebra, it is intolerable to have ⊃ used both for an operation between sentences ($p ⊃ q$) and for a relation between sets (A ⊃ B). The symbols in column 5 also are all traditionally associated with *relations* of various kinds and should not therefore be used for *operations*: usually they are favoured by authors who have not recognized the importance of distinguishing between these two concepts. This is discussed further later in this chapter when considering boolean relations, and again in Chapter 17 on implication.

The last column 6 of Table 16.1 gives various properties that are enjoyed by the 8 operations:

C commutative,   A associative,   Id idempotent.

**Table 16.1—Binary operations**

| 0 | 1 | 2 ACCEPTABLE SUBSTITUTES | | 3 | 4 | 5 | 6 |
|---|---|---|---|---|---|---|---|
| READING | PREFERRED NOTATION | SETS | LOGIC | ALTERNATIVE NOTATIONS | UNSUITABLE NOTATIONS | UNSUITABLE NOTATIONS | PROPERTIES OF OPERATION |
| or | $\vee$ | $\cup$ | | | $+$ | | C A Id |
| and | $\wedge$, $\cdot$ | $\cap$ | & | | | | C A Id |
| plus | $+$ | $\triangleleft$ | | $\oplus$, $\underline{\vee}$, $\sqcup$ | $\wedge$, $\div$ | $\Leftrightarrow$, $\equiv$, $=$, $\sim$ | C A |
| if and only if | $\leftrightarrow$ | $\triangleright$ | $\leftrightarrow$ | | $\times$ | $\not\equiv$ | C A |
| nor | $\rightarrow$ | | | $\parallel$, $\bar{v}$, $\bar{\cup}$ | | | C |
| nand | $\leftarrow$ | | | $\mid$, $\bar{\wedge}$, $\bar{\cap}$ | | | C |
| only if | $\uparrow$ | | | | $\cap$ | $\not\Vee$, $\wedge$, $\sim$ | |
| but not | $\nrightarrow$ | $\diagup$ | | | $\not\supset$, $-$ | $\not\Uparrow$, $\not\wedge$, $\sim$ | |

In a preliminary column, readings are suggested for each symbol. These have been chosen to be 'snappy' and at least reasonably suggestive of the operation in *all three* concrete systems.

### 16.3

The binary operation ① will be called the *contrary*† of operation ○ if, for all $x, y$,

$$x \oplus y = (x \circ y)';$$

operation ② is the *dual* of ○ if

$$x \oplus y = (x' \circ y')'.$$

Similar definitions can be applied to functions:

$$f_1 \text{ is the } contrary \text{ of } f \quad \text{if} \quad f_1(x, y, z) = [f(x, y, z)]'$$
$$f_2 \text{ is the } \quad dual \quad \text{of } f \quad \text{if} \quad f_2(x, y, z) = [f(x', y', z')]'$$

for all $x, y, z$.

The 8 essentially distinct binary operations are shown diagramatically in Fig. 16.1, *both* members of a pair of non-commutative operations (← and →; ↤ and ↦) being given. [One reason for preferring ↦ to \ for *general* use as 'but not' is that the sloping sign is not used reversibly.]

In Fig. 16.1, binary operations that are *contrary* are placed at diametrally *opposite vertices* of the dekagon; operations that are *dual* are joined by a *vertical line*.‡ The 2 operations ↔ and + are both contrary and dual. [The symbol ↔ having been constructed as the 'union' of the 2 signs ← and →, it is appropriate that + should turn out to be the 'intersection' of the symbols ↤ and ↦. But since, of course,

$$x \leftrightarrow y = (x \leftarrow y) \wedge (x \rightarrow y) \quad \text{and} \quad x + y = (x \leftarrowtail y) \vee (x \rightarrowtail y),$$

such a mnemonic is rather misleading; the 'union' and 'intersection' operations really go with the 'wrong' signs.]

---

†Because $x \oplus y$ is the complement of $x \circ y$, it is very tempting to call operation ① the complement of operation ○, but there is a good reason for *not* doing this: see Section 16.12.
‡Pairs of operations at the same *horizontal* level (such as ↑ and v) have a property like $x \oplus y = x' \circ y'$.

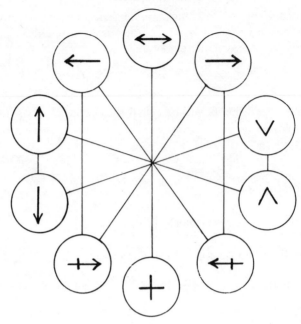

Fig 16.1—Binary operations

The dual operations ↓ and ↑ are the Sheffer† operations, each of which is capable of generating all possible boolean expressions. To discuss this fascinating phenomenon, however, would take us too far from our main theme. The reading 'nor' for ↓ (the negation of 'or') has suggested to circuit designers, who use these operations, the reading 'nand' for ↑ (the negation of 'and').

### 16.4

Having boldly claimed that there are 8 essentially distinct binary operations (or 10 if $x \circ y$ and $y \circ x$ are counted separately when $\circ$ is not commutative), it is perhaps incumbent on us to explain just why this is so. Concentrating first on a 2-valued boolean algebra, it is clear that $f(x, y)$ is determined once the values of $f(0, 0)$, $f(1, 0)$, $f(0, 1)$ and $f(1, 1)$ have all been selected. Since there are 2 choices (0 or 1) for each of these 4 quantities, there is a total of 16 ($= 2^4$) boolean functions of 2 variables. The identity

$$f(x, y) = [f(0, 0) \wedge x' \wedge y'] \vee [f(1, 0) \wedge x \wedge y'] \vee [f(0, 1) \wedge x' \wedge y]$$

$$\vee [f(1, 1) \wedge x \wedge y], \quad (1)$$

†Henry M. Sheffer (1883–1964).

or its dual form†

$$f(x, y) = [f(0, 0) \vee x \vee y] \wedge [f(1, 0) \vee x' \vee y] \wedge [f(0, 1) \vee x \vee y']$$
$$\wedge [f(1, 1) \vee x' \vee y'], \quad (2)$$

is the algebraic expression of this fact.

The 16 functions are listed systematically in Table 16.2 and it is seen that, when the 2 constant functions ($f(x, y) = 0$ and 1) and the 4 singulary functions ($f(x, y) = x, y, x'$ and $y'$) are discarded, there are exactly 10 that correspond to *genuinely* binary operations.

The column at the right of Table 16.2 gives the polynomial representation of each of the 16 expressions $f(x, y)$. This representation is discussed in Section 15.7. There are, of course, 16 and only 16

Table 16.2—Binary functions

| $f(0, 0)$ | $f(1, 0)$ | $f(0, 1)$ | $f(1, 1)$ | $f(x, y)$ | *Polynomial* |
|---|---|---|---|---|---|
| 0 | 0 | 0 | 0 | 0 | 0 |
| 1 | 0 | 0 | 0 | $x \downarrow y = x' \wedge y'$ | $1 + x + y + xy$ |
| 0 | 1 | 0 | 0 | $x \nrightarrow y = x \wedge y'$ | $x \quad + xy$ |
| 0 | 0 | 1 | 0 | $x \nleftarrow y = x' \wedge y$ | $y + xy$ |
| 0 | 0 | 0 | 1 | $x \wedge y$ | $xy$ |
| 1 | 1 | 0 | 0 | $y'$ | $1 \quad + y$ |
| 1 | 0 | 1 | 0 | $x'$ | $1 + x$ |
| 1 | 0 | 0 | 1 | $x \leftrightarrow y$ | $1 + x + y$ |
| 0 | 1 | 1 | 0 | $x + y$ | $x + y$ |
| 0 | 1 | 0 | 1 | $x$ | $x$ |
| 0 | 0 | 1 | 1 | $y$ | $y$ |
| 1 | 1 | 1 | 0 | $x \uparrow y = x' \vee y'$ | $1 \quad + xy$ |
| 1 | 1 | 0 | 1 | $x \leftarrow y = x \vee y'$ | $1 \quad + y + xy$ |
| 1 | 0 | 1 | 1 | $x \rightarrow y = x' \vee y$ | $1 + x \quad + xy$ |
| 0 | 1 | 1 | 1 | $x \vee y$ | $x + y + xy$ |
| 1 | 1 | 1 | 1 | 1 | 1 |

†These 2 identities (or their generalizations) give the so-called *normal* or *canonical* forms of expressions; the first is called the disjunctive and the second the conjunctive normal form.

boolean polynomials of just 2 variables, $x, y$, namely

$$\kappa + \lambda x + \mu y + \nu xy \quad \text{where} \quad \kappa, \lambda, \mu, \nu = 0 \text{ or } 1.$$

[Remember that $z + z = 0$ and $z \cdot z = z$.] There is a bijection between these 16 polynomials and the 16 sets of values $f(x, y)$ that define the function.

### 16.5

With set algebra, where the elements are *any* subsets of a given universal set I, it is perhaps not so obvious that the same considerations apply. The problem is to decide how many sets can be constructed from 2 given sets A and B using the operations $\{\cap, \cup, '\}$ of the algebra. But the above identities (1) and (2) still hold: reverting for the moment to a more familiar notation for sets, identity (1) says that

$$f(A, B) = [f(\varnothing, \varnothing) \cap A' \cap B'] \cup [f(I, \varnothing) \cap A \cap B'] \cup [f(\varnothing, I) \cap A' \cap B]$$
$$\cup [f(I, I) \cap A \cap B] \quad (3)$$

and, since any set constructed from $\varnothing$ and I by means of the operations $\{\cap, \cup, '\}$ must be either $\varnothing$ or I, it follows that, once it has been specified whether $f(\varnothing, \varnothing)$, $f(I, \varnothing)$, $f(\varnothing, I)$, $f(I, I)$ are to be $\varnothing$ or I in each case, the function $f$ is determined.

Indeed, although 'truth tables' and 'closure tables' tend to be studied only in reference to the 2-valued boolean systems (sentence logic and switching algebra), they *can* be introduced no less successfully into set algebra and made a feature of the exposition there. Any boolean operation between sets can be defined by what may be called an *admission table*, resembling a truth table; for example,

A ↛ B
(A \ B)

| A \ B | $\varnothing$ | I |
|---|---|---|
| $\varnothing$ | $\varnothing$ | $\varnothing$ |
| I | I | $\varnothing$ |

or, if arranged linearly,

| A | $\varnothing$ | I | $\varnothing$ | I |
|---|---|---|---|---|
| B | $\varnothing$ | $\varnothing$ | I | I |
| A↑B | I | I | I | $\varnothing$ |

.

Each such table is characteristic of one particular boolean set function. Observe the correlation between these admission tables and Table 16.2. The above identity (3) and these tables have an interpretation in terms of the conventional Venn diagram (Fig. 16.2). Two given subsets A, B of the universal set I, not specially related, divide the region representing I into 4 disjoint subregions and the set

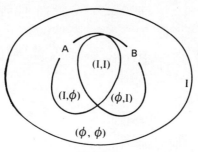

Fig 16.2

defined by any boolean expression built up from A and B can only admit or reject each of these 4 subregions *in its entirety*. The 16 functions correspond to the 16 ways in which such selections can be made. In fact, you will observe that the region marked $(\alpha, \beta)$ in Fig. 16.2, where $\alpha, \beta = \varnothing$ or I, will be admitted to[†] $f(A, B)$ if and only if $f(\alpha, \beta) = I$ in the admission table just mentioned.

### 16.6

There is another way in which a tabular presentation of set algebra can be achieved. This involves putting the emphasis on the actual *elements* $t$ of the universal set I. Every subset A of I has a *characteristic function*, $v_A$, say. This is the mapping

$$I \rightarrow \{0, 1\} : v_A$$

such that

$$v_A(t) = 1 \text{ if } t \in A \quad \text{and} \quad v_A(t) = 0 \text{ if } t \notin A.$$

[In the next chapter, this function $v_A$ will also be written $v[A]$, but for the present purpose that would be confusing.]

If ○ is any binary operation between sets, the characteristic function of A ○ B, $v_{A○B}$, is determined once its value $v_{A○B}(t)$ has been fixed for each of the 4 possible value pairs for $v_A(t)$, $v_B(t)$; for example,

|  |  | $v_B(t)$ | |
|---|---|---|---|
| $v_A(t)$ |  | 0 | 1 |
| $v_{A \rightarrow B}(t)$ | 0 | 1 | 1 |
|  | 1 | 0 | 1 |

[†]We are avoiding saying 'be included in', 'be contained in' or 'belong to' because of the other mathematical associations these phrases have.

Such tables can be used to define corresponding operations between the characteristic functions themselves, by writing

$$v_{A\circ}(t) \quad \text{as} \quad v_A(t) \circ v_B(t) \quad \text{or as} \quad (v_A \circ v_B)(t),$$

so that, more shortly,

$$v_A \circ v_B \quad \text{is} \quad v_{A\circ B},$$

where, for economy, the *same* symbol $\circ$ is being used for the operation between the functions. (But think carefully about what is being said; it is this sort of economy that can easily obscure delicate logical nuances.) So, dropping the $t$, such a table takes the form, for example,

| $v_A$ | 0 | 1 | 0 | 1 |
|---|---|---|---|---|
| $v_B$ | 0 | 0 | 1 | 1 |
| $v_A \leftrightarrow v_B$ | 1 | 0 | 0 | 1 |

Although having a slightly different notation and interpretation, these tables are isomorphic to the earlier admission tables (and the entries in Table 16.2 are exactly reproduced if $x$ is written for $v_A$ and $y$ for $v_B$). If you prefer, these tables can now be alternatively called *membership tables*, because they specify how the membership of A $\circ$ B is to be determined in terms of the membership of A and of B, 0 and 1 being associated with 'out' and 'in' respectively—a happy linguistic accident for English speakers![†]

### 16.7

It has just been said that the operation $\circ$ between the *functions* $(v_A \circ v_B)$ has been determined by the corresponding operation $\circ$ between the *sets* (A $\circ$ B). Historically, this may be a fair way of looking at matters, since the abstract algebra arose as a distillation of the various experiences with practical systems.

But *logically* it is more accurate to stand this claim on its head and say that it is the operation $\circ$ between the *sets* (A $\circ$ B) that is defined from the corresponding operation $\circ$ between the *functions* $(v_A \circ v_B)$.

Consider even a very basic binary operation, such as intersection. The set A $\cap$ B is defined by making

$$t \in A \cap B \quad \text{equivalent to} \quad t \in A \text{ and } t \in B.$$

[†]Membership tables are mentioned by Ian Stewart in *Concepts of modern mathematics* (Penguin).

This definition is perfectly clear: certainly, there is no suggestion that it is in any way inadequate for educational purposes.

But, when you come to analyse it, you will see that the definition is really saying that a particular one of the 16 abstract binary operations is being selected from Table 16.2—the one, in fact, denoted by '∧'—and the set A ∩ B is then specified by the requirement that

$$t \in A \cap B \quad \text{is equivalent to} \quad v_A(t) \wedge v_B(t) = 1,$$

that is, the set A ∩ B is defined by making

$$v_{A \cap B}(t) \quad \text{the same as} \quad v_A(t) \wedge v_B(t).$$

|  | $t \notin A$ $t \notin B$ | $t \in A$ $t \notin B$ | $t \notin A$ $t \in B$ | $t \in A$ $t \in B$ |
|---|---|---|---|---|
| $x \ = \ v_A(t)$ | 0 | 1 | 0 | 1 |
| $y \ = \ v_B(t)$ | 0 | 0 | 1 | 1 |
| $x \wedge y = v_A(t) \wedge v_B(t)$ | 0 | 0 | 0 | 1 |
|  | $t \notin A \cap B$ | $t \notin A \cap B$ | $t \notin A \cap B$ | $t \in A \cap B$ |

An exactly similar idea is at the heart of the definitions of A + B, A ← B and so on. Such sets (and hence the operations that generate them) are really determined by associating them in this way with the appropriate functions of $x = v_A(t)$, $y = v_B(t)$ conjured from Table 16.2. Viewed in the strictest and most fundamental way, it is the abstract operations that induce the set operations, not the other way round.

In the next chapter, an analogous procedure will be used to disentangle some knotty problems arising from the way elementary logic is presented, but it will be convenient to return now—after a brief digression—to more general observations.

## 16.8

You may have been wondering why, in Table 16.2 and elsewhere, the sequence of ordered pairs $(x, y)$ has been taken in the order 00, 10, 01, 11, rather than in ascending binary numeric order: 00, 01, 10, 11. There is a particular reason for this, but since it is in no way relevant to the present discussion, you will lose nothing by ignoring it. For those that are interested, however, the reason is that, when it comes to listing a table of values (a closure table, a truth table, an admission table) of an expression containing 3 or more variables, a binary scale ordering of the variables, such as

000, 001, 010, 011, 100, 101, 110, 111,

is not at all convenient or even appropriate. It is much more useful to group together all the ordered triples with the *same* number of unit entries and then, within each group, to give the positions lexical preference from, say, left to right. So, for 3 variables, this better sequence of ordered triples is

000; 100, 010, 001; 110, 101, 011; 111.

There is nothing strange about this. If you are asked to list the subsets of a set $\{\alpha, \beta, \gamma\}$ with 3 elements, you will probably present them in the order

$\varnothing, \{\alpha\}, \{\beta\}, \{\gamma\}, \{\alpha, \beta\}, \{\alpha, \gamma\}, \{\beta, \gamma\}, \{\alpha, \beta, \gamma\}.$

You are rather *un*likely to give them in the order

$\varnothing, \{\gamma\}, \{\beta\}, \{\beta, \gamma\}, \{\alpha\}, \{\alpha, \gamma\}, \{\alpha, \beta\}, \{\alpha, \beta, \gamma\},$

no matter how anxious you are to show off your familiarity with the binary system! Such an ordering is irrelevant here—just as it is in all similar catalogues.

For 4 variables, the corresponding sequence of ordered quadruples is

0000; 1000, 0100, 0010, 0001; 1100, 1010, 1001,

0110, 0101, 0011; 1110, 1101, 1011, 0111; 1111.

You will observe that the symmetry properties of the ascending binary numbers are not lost in these new sequences. But the latter have considerable additional advantages *not* enjoyed by the consecutive binary ordering usually used. Although these considerations are unimportant when expressions with only 2 variables are involved (and anyway a right over left preference could be used to make the sequences of *pairs* agree fortuitously), this writer prefers to stick to a sound general habit once acquired. The bonus then comes, for example, in Table 16.2 where, you will notice, this *same* ordering principle is applied to the *sets of values* that define the 16 functions. The use of an ascending binary scale there would have listed the functions in an even more muddled order, a fact you can readily confirm by consulting a book whose author has failed to notice the irrelevance and inconvenience of a binary ordering for the task in hand.

**16.9**

The discussion will now turn from boolean operations to boolean relations. Each binary function $f$ defines two binary relations by imposing the requirement

either that $f(x, y)$ only takes the value 1

or    that $f(x, y)$ only takes the value 0.

These conditions will be represented by '$f(x, y) \equiv 1$' and '$f(x, y) \equiv 0$' respectively. (They can be read as '$f(x, y)$ must be 1', etc.) One does not want to write such a condition as merely '$f(x, y) = 1$', because it is convenient elsewhere to be able to make statements like '$f(x, y) = 1$ when $x = 1$, $y = 0$'. To try to use the same sign '=' with both interpretations would mean either giving up this freedom or risking ambiguity.

Since '$f(x, y) \equiv 0$' is equivalent to '$f_1(x, y) \equiv 1$', where $f_1$ is the contrary of $f$, no relations are lost by ignoring the second possibility. The 16 binary functions therefore give rise to 16 distinct binary relations. (Those associated with the two constant functions are, of course, the empty relation and the universal relation.)

The relations corresponding to the 8 genuinely binary operations are listed in Table 16.3, with their definitions in column 1. Just as Table 16.1 gave a list of some relevant properties of the 8 *operations* (commutative, associative, idempotent), so the last column 6 of Table 16.3 includes observations on some relevant properties of the 8 *relations*:

 R reflexive, Ir irreflexive, S symmetric, An antisymmetric, T transitive.

A comparison will help to reinforce in your mind the discussion in Chapter 1 on the need to make sure that any questions asked are appropriate to the concept.


**16.10**

With the boolean operations listed in Table 16.1, a reading was suggested that could sensibly be used in all three of the concrete systems; a convergence towards some such standard nomenclature would certainly be beneficial for the subject. It is, however, difficult to choose a single reading for most of the boolean relations, because the possible linguistic descriptions of the relations tend to be dependent on the physical nature of the elements in the structure. But fortunately this is

## Table 16.3—Binary relations

| 0 RELATION | 1 DEFINING EQUATION | 2 FANCIFUL GENERAL NOTATION | 3 SWITCHING ALGEBRA $x = v[a]$, $y = v[b]$ | 4 SENTENCE LOGIC $x = v[p]$, $y = v[q]$ | 5 SET ALGEBRA $x = v[A]$, $y = v[B]$ | 6 PROPERTIES OF RELATION |
|---|---|---|---|---|---|---|
| alternation | $x \vee y \equiv 1$ $(x \downarrow y \equiv 0)$ | $x \vee y$ | a and b are alternative $(a' \leqslant b)$ | p and q are exhaustive $(p' \Rightarrow q)$ | A and B cover I $(A' \subseteq B)$ | S |
| acceptance | $x \wedge y \equiv 1$ $(x \uparrow y \equiv 0)$ | $x \wedge y$ | a and b are both closed $(a > b')$ | p and q are both true | $A = I$ and $B = I$ | S T |
| complementarity | $x + y \equiv 1$ $(x \leftrightarrow y \equiv 0)$ | $x \nleftrightarrow y$ | a is equivalent to the complement of b $(a \Leftrightarrow b';\ a = b')$ | p contradicts q — p is equivalent to the negation of q $(p \Leftrightarrow q')$ | A is the complement of B $(A = B')$ | Ir S |
| equivalence | $x \leftrightarrow y \equiv 1$ $(x + y \equiv 0)$ | $x \leftrightarrow y$ | a is equivalent to b $(a \Leftrightarrow b;\ a = b)$ | p implies and follows from q; p is equivalent to q $(p \Leftrightarrow q)$ | A is equal to B $(A = B)$ | R S T |
| rejection | $x \downarrow y \equiv 1$ $(x \vee y \equiv 0)$ | $x \downarrow y$ | a and b are both open $(a' > b)$ | p and q are both false | $A = \varnothing$ and $B = \varnothing$ | S T |
| incompatibility | $x \uparrow y \equiv 1$ $(x \wedge y \equiv 0)$ | $x \uparrow y$ | a and b are incompatible $(a \leqslant b')$ | p and q are mutually exclusive $(p \Rightarrow q')$ | A and B are disjoint $(A \subseteq B')$ | S |
| inclusion (implication) | $x \rightarrow y \equiv 1$ $(x \nrightarrow y \equiv 0)$ | $x \Rightarrow y$ | a is included in b $(a \leqslant b)$ | p implies q $(p \Rightarrow q)$ | A is contained in B $(A \subseteq B)$ | R An T |
| discrimination | $x \nrightarrow y \equiv 1$ $(x \rightarrow y \equiv 0)$ | $x \nleftrightarrow y$ | a is closed but b is open $(a > b)$ | p is true but q is false | $A = I$ and $B = \varnothing$ | Ir |

not too important. Boolean relations are inevitably studied piecemeal and obviously many of the 8 possible relations are normally ignored. The reason for presenting a conspectus here is that no book known to the writer explicitly draws the reader's attention to the complete set of binary relations that are theoretically possible and their connexion with the set of operations.

Indeed, many authors cause untold confusion by muddled definitions that blur the distinction between operations and relations and by the use of symbols and names traditionally associated with *relations* (signs such as $\leq$, $\not\equiv$ and others listed in column 5 of Table 16.1 and titles such as 'inclusion' and 'non-equivalence') for their *operations*! The aim of this chapter is to help you to chart a passage through these treacherous channels so that, by clarifying your own ideas, you will be better able to explain important concepts to your pupils, even though the details of what is being said here will probably lie well beyond their immediate interests.

Because of the difficulty of finding single readings for boolean relations that will be appropriate to all contexts in which they occur, Table 16.3 has moved in the opposite direction and avoided giving undue prominence to any one reading by emphasizing in columns 3, 4, 5 the variety of possible descriptions that a single (abstract) relation may enjoy. This concordance may draw your attention to links you had not previously forged for yourself. It is left to you to reduce this rich variety of terminology to a more manageable, standard list if you judge this task to be possible. (Please ignore for the present the headings to these columns, $x = v[a]$, $y = v[b]$, etc.)

## 16.11

Most of the boolean relations do not have any recognized symbolic representation at all, and, for those that do, there is a variety of notation, arising from the diversity of readings just mentioned.

Perhaps the most interesting binary relation is the partial order defined by '$x \rightarrow y \equiv 1$'. In logic, this corresponds to the implication relation '$p \Rightarrow q$' (which is considered in the next chapter) and in set algebra to the inclusion relation '$A \subseteq B$'. The *set* $A \rightarrow B$ is defined as $A' \cup B$ and, when constructed from general subsets A and B of I, it is represented by the region of the Venn diagram that excludes the shaded area (Fig. 16.3). The corresponding relation '$A \rightarrow B \equiv I$' says that '$A \rightarrow B$ is the universal set', that is, that the shaded area is *empty* and that the sets are related by inclusion: $A \subseteq B$. In switching algebra, $a \rightarrow b$, defined as $a' \vee b$, denotes a particular *switch*, constructed by connecting an $a'$ switch and a $b$ switch in parallel (Fig. 16.4). If $a$ and $b$ are independent

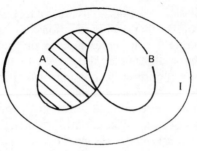

Fig 16.3

switching variables, then $a \to b$ is open for certain values of $a$ and $b$ (in fact just for $a = 1$, $b = 0$) and closed for the others. On the other hand, '$a \to b \equiv 1$' says that '$a \to b$ is a closed switch in all circumstances'. This means that the switches $a$ and $b$ are related and do *not* operate independently. It is possible to have

$$\left.\begin{array}{ll} a = 0, & b = 0, \\ a = 0, & b = 1, \\ a = 1, & b = 1, \end{array}\right\} \quad \text{but } not \quad a = 1, \quad b = 0.$$

In this subject, this partial order relation is again called inclusion, to indicate that the circuit-closing capacity of $a$ is included in that of $b$, but is here usually symbolized by '$a \leqslant b$'.

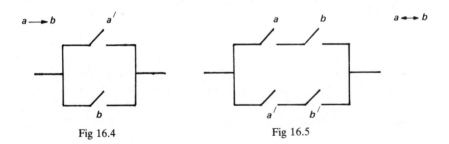

Fig 16.4                    Fig 16.5

The relation determined by '$x \leftrightarrow y \equiv 1$' is an equivalence. A switch can similarly be defined based on '$\leftrightarrow$'; it is just one version of the familiar 'staircase' switch:† a pair of 2-way (single pole, double throw) switches connected as in Fig. 16.5. The switch $a \leftrightarrow b$ is closed when $a$

†The other version models the operation $+$: compare the remarks in Section 16.1.

and $b$ are both closed and also when they are both open, but is otherwise open. But the relation defined by '$a \leftrightarrow b \equiv 1$' ($a \leftrightarrow b$ must be closed) says that $a$ and $b$ always operate simultaneously and so are equivalent switches. The best way of writing this is '$a \Leftrightarrow b$', although '$a = b$' is, in fact, more usual. (The merits of this are discussed in Section 17.20.) If you draw for yourself a diagram representing the set $A \leftrightarrow B$ (it is the complement of $A + B$), you will observe—and easily prove—that '$A \leftrightarrow B \equiv I$' is the relation '$A = B$' (and here the sign '$=$' can be used without ambiguity). In logic, on the other hand, such a notation is quite unsuitable: the equivalence of the propositions $p$ and $q$ (the statement that $p$ and $q$ are either both false or both true) is denoted by '$p \Leftrightarrow q$'. These observations also are expanded in Chapter 17.

## 16.12

In order to be able to present certain arguments more forcefully, a full set of reasonable symbols has been devised for the 8 relations in column 2 of Table 16.3 and is offered—not too seriously—for use in all the systems, but particularly in the abstract boolean algebra. The signs '$\Rightarrow$' and '$\Leftrightarrow$', although familiar in logical contexts, are scarcely used in the other practical systems, but they are convenient for the abstract applications envisaged. The other symbols have been constructed in a similar sort of way by duplicating one of the main structural elements in the operation sign.

It must be confessed that the main purpose of this whimsical exercise is to reinforce the discussion in Chapter 1 about the distinction in concept between a relation and an operation, and to reiterate the verbal nature of a relation symbol by providing such a mark to stand in the place of $\rho$ in a statement such as '$x \, \rho \, y$' whenever required.

But a further reason for inventing these fanciful general purpose symbols is that the relations can then be displayed in a diagram, Fig. 16.6, which resembles Fig. 16.1 for operations, and enables an important, but rather subtle, point to be made.

Following the pattern set in Section 16.3 for operations, it is reasonable to call relations (such as $\vee$ and $\Downarrow$) situated at diametrically opposite vertices of the dekagon *contrary relations* and those (like $\vee$ and $\wedge$) joined by a vertical line *dual relations*. (Then, if $\rho_1$ is the contrary of $\rho$ and $\rho_2$ the dual of $\rho$, $x \, \rho_2 \, y \Leftrightarrow x' \, \rho_1 \, y'$: but a definition of either $\rho_1$ or $\rho_2$ in terms of $\rho$ requires a mention of the associated contrary or dual *operation*.)

The fact that needs stressing, because it can so easily lead to misunderstanding, is that there is a distinction between contrary boolean relations and complementary relations. You will recall from Chapter 1

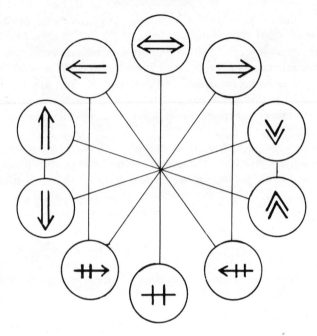

Fig 16.6—Binary relations

that the relation $\rho'$ is called the *complement* of relation $\rho$ if

$$(x, y) \in \Gamma_{\rho'} \quad \Leftrightarrow \quad (x, y) \notin \Gamma_{\rho},$$

where $\Gamma_{\rho}$, $\Gamma_{\rho'}$ are the graphs of $\rho$ and $\rho'$.

Now, for example, '$x \veebar y$' says that '$x \vee y \equiv 1$': that is, in the case of a 2-valued algebra, that the pair of values $x = 0$ and $y = 0$ are *never* taken simultaneously; in the case of set algebra, that the region outside both the given sets A and B is empty. The *complement* of that relation says that the values $x = 0$ and $y = 0$ *do* occur simultaneously (but leaves quite open the question of what other values, if any, may *also* occur); in the case of sets, that there is at least one element of I that belongs to neither of the given sets. But the *contrary* of this relation, $x \Downarrow y$, gives the information that '$x \downarrow y \equiv 1$' and hence that $x = 0$ and $y = 0$ (A $= \varnothing$ and B $= \varnothing$) *must* occur: that is the *only* possibility this relation allows.

Similarly, '$x \Leftrightarrow y$' says that '$x \leftrightarrow y \equiv 1$': that is, in the case of a 2-valued algebra, that the values taken by $x$ and $y$ must be equal; in the case of sets, that the given sets A and B are identical (A $=$ B). The *complementary* relation to this ($x$ is *not* equivalent to $y$, $x \not\Leftrightarrow y$) asserts that the values taken by $x$ and $y$ are *not always equal* (although they may sometimes be), that is, that there are definitely circumstances in which the value of $x$ is different from the value of $y$; for sets, that there is at least one element of I that belongs to one of the sets A, B but *not* to the other $(A \neq B)$. The *contrary* relation, however, here denoted by '$x \mathbin{+\!\!\!\!\!+} y$', says that '$x + y \equiv 1$', so that, in a 2-valued algebra, $x$ is *always different* from $y$ (if $x = 0$, then $y = 1$ and, if $x = 1$, then $y = 0$); with sets, that set A is the complement of set B. So $\mathbin{+\!\!\!\!\!+}$ is clearly different in meaning from $\Leftrightarrow$ (which latter relation symbol is likely, in certain contexts, to be written $\neq$).

[Those of you who recognize the more formal type of symbolism below will appreciate that, if

$$x \,\rho\, y \quad \text{says} \quad \forall x \, \forall y \, P(x, y),$$

then

$$x \,\rho'\, y \quad \text{says} \quad \rightarrow\{\forall x \, \forall y \, P(x, y)\},$$

$$x \,\rho_1 y \quad \text{says} \quad \forall x \, \forall y\{\rightarrow P(x, y)\}.]$$

With every binary boolean relation there is a similar distinction to be detected between its contrary and its complement. So, even though the definition of a binary operation ① contrary to operation ○ involves nothing more than the simple taking of a complement,

$$x \oplus y = (x \circ y)',$$

operations such as ○ and ① are better called contrary (and *not* complementary), because the associated (contrary) relations defined by

$$x \circ y \equiv 1 \quad \text{and} \quad x \oplus y \equiv 1$$

are *not* complementary relations.

The word 'complementary' does tend to be overworked. Yet each of the underlying definitions here (complementary sets; complementary relations) involves quite a sensible and logical use of the term. Indeed, the ideas are closely connected: complementary relations have for their graphs complementary subsets of the relevant cartesian product. It is only when the two ideas come together in this aspect of the theory that they clash and here care is needed.

# Chapter 17

# Implication

---

## 17.1

The discussion in Chapter 1 on the distinction between an operation and a relation will, it is hoped, help you to construct appropriate explanations of these ideas for your own pupils. Many textbook writers now understand the general issues involved quite well and provide satisfactory treatments. But, in the present writer's opinion, almost every author that tries to explain sentence logic (the propositional calculus) at an *elementary* level comes to grief when he talks about implication and writes nonsense by professing that it has a truth table. The claim is made in this book that implication is a relation and so it is as meaningless to say that '$p$ implies $q$' has a truth table as it is to say that '$t > u$' has a numeric value or that '$l \parallel m$' is a line. This bold contradiction of what many eminent authorities have written will take some justification. The argument will be developed slowly and steadily, in order to persuade you that the case is sound. If you are not convinced, one can only hope that you will have found the discussion stimulating.

The starting points for the reader are not only Chapter 1 but also the material on boolean operations and relations in Chapter 16. Many relevant preliminary observations are included in those chapters, although with only passing reference to the specific problem now to be elucidated. It should be pointed out that some of the remarks early in this chapter about sentence variables and truth tables (which artlessly copy those in many conventional presentations) are themselves going to come under critical examination later (for a different reason) and it will then be found that these notions will benefit from a more careful scrutiny than they are given below.

Since this chapter is predominantly concerned with logical considerations, the notation to be used for the (sentence) variables will be $p$, $q$, $r$, $s$, with either 0 or F used for 'false' and either 1 or T for 'true'. In order to keep the letters $x, y, z$ for use as elements of the abstract boolean algebra, the real numbers that occur in various illustrative propositions will be denoted by $t$ and $u$.

It has already been explained (see Table 16.2) that there are 16 and only 16 distinct expressions $f(p, q)$ involving 2 variables, each having a truth table that specifies the choice to be made for each of the 4 entries

$$f(0, 0), \quad f(1, 0), \quad f(0, 1), \quad f(1, 1).$$

In the context of logic, operations (particularly binary operations) are often called *propositional connectives* or just *connectives*. This is an absolutely splendid term, which vividly emphasizes their grammatic role. (It is a term that could perhaps be employed, without disadvantage, in other areas of mathematics.)

## 17.2

The particular connective denoted by '$\rightarrow$' is an important operation in logic; it is defined by the truth table

| $p$ | 0 | 1 | 0 | 1 |
|---|---|---|---|---|
| $q$ | 0 | 0 | 1 | 1 |
| $p \rightarrow q$ | 1 | 0 | 1 | 1 |

or, if this looks more familiar,

| $p$ | F | T | F | T |
|---|---|---|---|---|
| $q$ | F | F | T | T |
| $p \rightarrow q$ | T | F | T | T |

.

In other words, the definition makes '$p \rightarrow q$' identical to '$p' \vee q$'. Most people read the compound sentence '$p \rightarrow q$' as 'if $p$, then $q$', but a much *better* reading is surely '$p$ only if $q$', since this puts all the words in the place where the infix symbol is.

The operation $\rightarrow$ is non-associative, because

$$(0 \rightarrow 0) \rightarrow 0 \ = 1 \rightarrow 0 = 0$$
$$0 \rightarrow (0 \ \rightarrow 0) = 0 \rightarrow 1 = 1,$$

and so careful bracketing is necessary.

The operation $\rightarrow$ is clearly also non-commutative. The reversed operation $\leftarrow$ is defined so that '$p \leftarrow q$' is the same as '$q \rightarrow p$' and has the truth table

| $p$ | 0 | 1 | 0 | 1 |
|---|---|---|---|---|
| $q$ | 0 | 0 | 1 | 1 |
| $p \leftarrow q$ | 1 | 1 | 0 | 1 |

.

The sentence '$p \leftarrow q$' is interchangeable with '$p \vee q$''; '$p \leftarrow q$' is read as '$p$ if $q$'. The readings used remind one of the idea of conditionality inherent in these operations and, in fact, $\rightarrow$ and $\leftarrow$ are called *conditional connectives*.

**17.3**

Many teachers introduce this subject by saying 'Well; those are the truth tables for "and" and for "or". What we are going to look at next are the truth tables for "if" and for "only if"'. In consequence, the student often has difficulty with the definition of the connective $\rightarrow$ and he expresses his bewilderment by asking why '$p$ only if $q$' is required to be true when $p$ is false. The pupil is really asking the wrong question, but teachers usually struggle to convince him of the reasonableness of the mathematicians' definition by producing contrived colloquial examples of compound sentences of the type '$p$ only if $q$' in which $p$ is false (with both choices for $q$) and bludgeoning him into accepting that '$p$ only if $q$' '*is*' true in those cases. (The writer blushes to recall that he has himself been guilty of this tactic in his youth.) As so often happens with such dilemmas, however, the difficulty vanishes as soon as the problem is viewed from the correct standpoint.

It is *not* the case that mathematicians have engaged in life-or-death struggles to determine at all costs the 'correct' truth table for '$p$ only if $q$' [or for 'if $p$, then $q$', if you still feel happier with that reading]. What they have before them is a *particular* truth table corresponding to one of the 16 possible binary connectives and, because it is perceived (as you yourself will be reminded when reading this chapter) that this connective is going to have special importance for the development of the subject, they look around for a suitable way of *reading the symbol* informally. *Because* the definition of '$p \rightarrow q$' has made it equivalent to '$p' \vee q$', and *because*, in ordinary speech, '$p$ only if $q$' is often synonymous with 'not $p$, or $q$', it seems a sensible solution to *read* the *precisely-defined* proposition '$p \rightarrow q$' using the more *loosely-defined* phrase '$p$ only if $q$' and in so doing to clamp on to that slightly nebulous phrase the precision of the truth table definition. Viewed in this way, the informal colloquial examples mentioned earlier have real value in explaining why the *reading* 'only if' [or 'if . . ., then . . .'] for '$\rightarrow$' is not too unreasonable, even in cases where it is rarely used in ordinary speech. No longer, however, are those tortuous sentences being used to buttress a claim that '$p \rightarrow q$' '*must*' be given the value T when $p$ is false '*because*' the proposition '$p$ only if $q$' '*is*' true when $p$ is false.

If the student believes that *he* uses the phrase '$p$ only if $q$' differently and thinks this sentence should have been given a different truth value in a certain case, he is no longer accusing the experts of having chosen the 'wrong' truth table for their definition of 'only if'. Mathematicians know perfectly well, as he also will later, why this table is being singled out for attention. *It is the connective defined by the table that is important, not the way it is read.*

It ceases to be a question of the student saying: 'I think their definition was wrong; I think "*p* only if *q*" should have been assigned such-and-such a table' in the quite mistaken belief that there is some compelling reason for '*p* only if *q*' to be *given* an exhaustive tabular interpretation. No such compulsion exists.

It becomes instead merely a question of his saying: 'I don't think "*p* only if *q*" is a very clever way of *reading* "*p* → *q*"; if I had had my way, it would have been the connective defined by such-and-such a table that *I* should have read as "*p* only if *q*" '. And he is perfectly entitled to that opinion. If he wishes, he need not give *any* colloquial reading to '*p* → *q*'. (Indeed, there would be fewer misunderstandings if we *all* agreed to deny ourselves that luxury!) He can just say something like '*p* arrow *q*' instead (and presumably '*p* worra *q*' for '*p* ← *q*'!); or he may care to invent some distinctive way of reading it that satisfies him. None of his linguistic beliefs, however, will alter the fact that the connective → is defined by making '*p* → *q*' the same as '*p'* ∨ *q*' and in no other way, because it is the connective with that truth table that is required, however the squiggle '→' may be vocalized.

[This conundrum is merely another example of the problem one comes up against time and time again in mathematics. Faced with a new concept that is being given a precise definition, should one invent a new and artificial word specifically to describe it and risk the topic becoming jargon-ridden and forbidding to the learner,† or should one pick a familiar word from ordinary speech that has some vague similarity of meaning and graft on to that word a new and exact technical meaning with the consequent danger of confusion (as illustrated so eloquently above)? Sometimes mathematicians adopt one solution, sometimes the other. Whichever they do, it seems to cause headaches for the school teacher! Topology is perhaps the outstanding example of the first approach, with a liberal introduction of words like homotopy, diffeomorphism, cohomology, compactification and so on to daunt the neophyte; and mechanics the prime example of the latter, with colloquial words like force, pressure, work, energy, power, stress, strain and so on suddenly acquiring exact scientific meanings that often conflict sharply with the layman's usage.]

---

†The terminology associated with mappings and morphisms (see Sections 2.10 and 2.11) is one such potential disaster area.

**17.4**

So '$p \rightarrow q$' is a well-defined proposition and, when $p$ and $q$ are general propositions, not specially related, '$p \rightarrow q$' may take either truth value, 0 or 1, according to the truth values of $p$ and $q$.

The notation '$p \Rightarrow q$' can now be introduced to express a certain *relation* between the sentences $p$ and $q$; this means, in particular, that $p$ and $q$ are not independent. The statement

$$'p \;\Rightarrow\; q' \quad \text{means that} \quad 'p \rightarrow q \quad \text{is true'};†$$

that is, that '$p$ is false or $q$ is true'.

The reading for '$p \Rightarrow q$' will be '$p$ implies $q$'. Note that, just as in the case of the operation $\rightarrow$ and its articulation 'only if', the introduction of the reading 'implies' for the relation $\Rightarrow$ is being made merely for convenience in verbal communication. The interpretation and modus operandi of the symbol '$\Rightarrow$' are to be governed solely by its definition in the preceding paragraph and not by any linguistic associations that the informal reading 'implies'—or indeed any other reading—might conjure up in the reader's mind.

Since

$$0 \rightarrow 0 = 1 \quad \text{and} \quad 1 \rightarrow 1 = 1,$$

$$'p \rightarrow p' \quad \text{is true.}$$

Hence

$$p \;\Rightarrow\; p,$$

which says that the relation of implication is reflexive.

Since

$$0 \rightarrow 1 = 1 \quad \text{but} \quad 1 \rightarrow 0 = 0,$$

the truth of '$p \rightarrow q$' does not guarantee the truth of '$q \rightarrow p$' and so the relation of implication is *not* symmetric. (As you will recognize, the distinction between '$p \Rightarrow q$' and '$q \Rightarrow p$' is that between a theorem and its converse.)

---

†More precisely, ' "$p \rightarrow q$" is true', but such secondary quotation marks will usually be omitted in cases like this, when the meaning is felt to be clear. You may also have noticed that even primary quotation marks are being suppressed around *single* letters, such as '$p$', even though on many occasions strict protocol would require their insertion. Such extreme formality, however, seems unnecessary here.

On the other hand, the relation of implication is transitive. To establish this, one has to show that, if '$p \rightarrow q$' is true and '$q \rightarrow r$' is true, then '$p \rightarrow r$' is true.

Suppose then that '$p \rightarrow r$' is false. It follows that

$$p \text{ is true } \textit{and} \quad r \text{ is false.}$$

So

if $q$ is false, then '$p \rightarrow q$' is false and, if $q$ is true, then '$q \rightarrow r$' is false.

In either case, one of '$p \rightarrow q$' and '$q \rightarrow r$' is false. Thus

if '$p \rightarrow q$' is true *and* '$q \rightarrow r$' is true, then '$p \rightarrow r$' is true,

that is,

$$\text{if} \quad p \ \Rightarrow \ q \quad \text{and} \quad q \ \Rightarrow \ r, \quad \text{then} \quad p \ \Rightarrow \ r.$$

Notice how the questions that have been asked about the verb '$\Rightarrow$' (reflexive? symmetric? transitive?) were appropriate ones to ask about a relation; contrast these with the questions asked earlier about the connective '$\rightarrow$' (commutative? associative?).

## 17.5

In mathematical *reasoning*, it is the relation of implication ($\Rightarrow$) that is used, *not* the conditional connective ($\rightarrow$). For example, in a context in which $u$ is understood to be a real number, one writes

$$\text{'}u < -5 \ \Rightarrow \ u^2 > 25\text{'}.$$

The message being conveyed here is that it is impossible for '$u < -5$' to be true and '$u^2 > 25$' to be false simultaneously; that is, that either '$u < -5$' is false or '$u^2 > 25$' is true.

The sentence

$$\text{'}u < -5 \ \rightarrow \ u^2 > 25\text{'}$$

may, of course, be offered for consideration, just as one may similarly offer

$$\text{'}u < 3 \ \vee \ u > 7\text{'}.$$

But it makes equal *sense* to examine

$$`u < -5 \quad \rightarrow \quad u^2 > 40',$$

which is, by definition, a proposition ($`u \nless -5 \lor u^2 > 40'$) that is inherently capable of being either true or false. In fact, this particular sentence is true if $u = -7$, 6 or 7, but false if $u = -6$ and these happen to illustrate all 4 cases in the truth table.

|  | $u = 6$ | $u = -6$ | $u = 7$ | $u = -7$ |
|---|---|---|---|---|
| $u < -5$ | F | T | F | T |
| $u^2 > 40$ | F | F | T | T |
| $u < -5 \rightarrow u^2 > 40$ | T | F | T | T |

Precisely because the *possibility* of

$$`u < -5 \quad \rightarrow \quad u^2 > 40'$$

being false *can* occur, it is *invalid* to claim that $`u < -5$ implies $u^2 > 40'$ or to write $`u < -5 \Rightarrow u^2 > 40'$. Since $`u < -5'$ can be true and $`u^2 > 40'$ can be false simultaneously (for example, when $u = -6$), any such statement conflicts with the definition just made (and, for that matter, would conflict with colloquial usage as well). This point will be returned to later (Section 17.12).

On the other hand, since

$$`u < -5 \quad \rightarrow \quad u^2 > 25'$$

*cannot* be false for any real $u$, the claim

$$`u < -5 \quad \Rightarrow \quad u^2 > 25'$$

is valid. This says that there is a *relation* between the propositions $`u < -5'$ and $`u^2 > 25'$: either $`u < -5'$ is false or $`u^2 > 25'$ is true.

Similarly, the statement

$$`u < -5 \quad \Rightarrow \quad u^2 > 20',$$

although mathematically weaker, is also logically unassailable, since it asserts that

$$` \text{``} u < -5 \quad \rightarrow \quad u^2 > 20 \text{''} \quad \text{is true'}.$$

**17.6**

As a further example, this time from within logic itself, observe that both the sentences

$$'(p \vee q) \;\rightarrow\; (p \wedge q)'$$

and

$$'(p \wedge q) \;\rightarrow\; (p \vee q)'$$

are well-defined for all $p$ and $q$; they can be examined and their truth values determined. (Try this as an exercise.) You will recognize the second as a *tautology*; that is, it takes the truth value 1 for *all* assignments of truth values to $p$ and $q$. Hence it is legitimate to write

$$'p \wedge q \;\Rightarrow\; p \vee q'.$$

But the converse implication is *not* valid when $p$ and $q$ are general propositions, because the first of the above pair of sentences *can* be false. (Similarly, '$A \cap B \subseteq A \cup B$' is universally valid, but '$A \cup B \subseteq A \cap B$' is not.)

Logicians make a distinction (which is not being made in this book, because it would *not* be useful at the most elementary level), between *syntactic* and *semantic* implication. Syntactic implication arises as a consequence of the structure of the logic itself and can be illustrated by the result '$p \wedge q \Rightarrow p \vee q$' just mentioned. Syntactic implication does not depend on what the particular propositions $p$ and $q$ denote and (in sentence logic) can be established by an entirely mechanical decision procedure. Semantic implication, on the other hand, is an implication demanded by the properties of the theory whose results are being given formal expression; the result '$u < -5 \Rightarrow u^2 > 25$' is an example of semantic implication. But whether syntactic or semantic, implication is still a *relation* between two sentences.

**17.7**

If you are familiar with predicate logic (in which the structure of individual sentences is analysed), you may be smugly thinking that any crisis of understanding is imaginary and only arises because the discussion is taking place within the framework of sentence logic (in which sentences are treated as indivisible entities) and that the whole problem, if problem there be, will go away if only one splashes a few quantifiers around. Let us try this. Denote '$u < -5$' by $Q(u)$, '$u^2 > 25$' by $R(u)$ and '$u^2 > 40$' by $S(u)$.† The conventional way of writing

$$\text{'}u < -5 \text{ implies } u^2 > 25\text{'}$$

in predicate logic would be

$$\forall u\{Q(u) \rightarrow R(u)\}. \tag{1}$$

But, if '$Q(u) \rightarrow R(u)$' is a proposition defined by a truth table, (1) is devoid of meaning: it is not sensibly constructed. What (1) is intending to say is surely

$$\forall u\{\text{'}Q(u) \rightarrow R(u)\text{' is true}\},$$

in other words,

$$\forall u\{Q(u) \;\Rightarrow\; R(u)\}. \tag{2}$$

It is still the *assertion* of '$Q(u) \rightarrow R(u)$' that is being claimed, just as it was before the decorative quantifier was added. The statement

$$\forall u\{Q(u) \rightarrow S(u)\} \tag{3}$$

can be written, and it is no more—but equally no less—sensible than (1). If one writes

$$\forall u\{Q(u) \;\Rightarrow\; S(u)\}, \tag{4}$$

which, of course, one would not normally do, then one has made a mistake. But sentence (4) is properly constructed: it just happens to be false.

Analogies are often dangerous and perhaps the ones to be mentioned now are more dangerous than most. But one can rather vividly juxtapose these statements (1)—(4) and the following collection, in which the sentences $Q(u)$, $R(u)$, $S(u)$ are replaced by certain algebraic expressions (involving a real number $t$), the operation $\rightarrow$ of logic is replaced by the operation $-$ of arithmetic and the relation $\Rightarrow$ of logic by the relation $>$ of arithmetic.

$$\forall t(t^2 + 5 - 2t) \tag{1a}$$

---

†Notice that the predicate letters Q, R, S represent singulary *relations*, *not* (as the traditional notation tends to suggest) singulary functions.

$$\forall t(t^2 + 5 > 2t) \tag{2a}$$

$$\forall t(t^2 + 5 - 6t) \tag{3a}$$

$$\forall t(t^2 + 5 > 6t) \tag{4a}$$

Like (1) and (3), (1a) and (3a) are meaningless, because improperly constructed (without a main verb); (2a) and (4a) are well formed and both can be examined for their falsity or truth: (2a) like (2) is true; (4a) like (4) is false.

A second analogy, drawn from set algebra, may be even more persuasive.

$$\forall A \; \forall B\{(A \cap B) \to B \} \tag{1b}$$

$$\forall A \; \forall B\{ \; A \cap B \; \subseteq B \} \tag{2b}$$

$$\forall A \; \forall B\{(A \cap B) \to B'\} \tag{3b}$$

$$\forall A \; \forall B\{ \; A \cap B \; \subseteq B'\} \tag{4b}$$

These again match (1)–(4) exactly.

Both analogies are imperfect because the 'internal' verb in $Q(u)$ has no counterpart in either the algebraic expression $t^2 + 5$ or the set $A \cap B$, but they seem to be close enough to make clear the points at issue.

With existential quantifiers, no less than with universal ones, the same distinctions can be discerned. Neither can be applied to a sentence until that sentence has been asserted. The two quantifiers are not, of course, independent concepts, because

$$\exists t \; P(t) \quad \text{is the same as} \quad \to (\forall t\{\to P(t)\}),$$

$$\forall t \; P(t) \quad \text{is the same as} \quad \to (\exists t\{\to P(t)\}),$$

where the negation of $p$ has here been denoted by $\to p$, rather than $p'$.

The negation of '$Q(u) \to S(u)$' is '$Q(u) \nrightarrow S(u)$', but the sentence

$$\exists u\{Q(u) \nrightarrow S(u)\} \tag{5}$$

is just as defective in its construction as example (3) was earlier. On the other hand, the sentence

$$\exists u\{\text{'}Q(u) \nrightarrow S(u)\text{'} \text{ is true}\},$$

which, using the notation of Chapter 16, becomes

$$\exists u\{Q(u) \nrightarrow S(u)\}, \tag{6}$$

is *correctly* formed and corresponds to statement (4). When Q and S have the meanings they were given earlier, (6) is seen to be true, since there does exist a value of $u$ (for example, $u = -6$) for which at the same time $Q(u)$ is true and $S(u)$ is false. Note that (6) is the negation of (4) and that the relation ($\not\rightarrow$) occurring in (6) is, as expected, the contrary of the relation ($\Rightarrow$) occurring in (4), *not* its complement ($\not\Rightarrow$). [See Section 16.12.]

### 17.8

If you agree with what has been written so far, you will see that it is nonsense to say that the connective here being written '$\rightarrow$' corresponds to the concept 'implies'—no operation can be a verb—and equally absurd to say that the relation here expressed by '$\Rightarrow$' has a truth table. Yet the lack of discrimination between these two ideas, here contrasted by the use of different symbols, seems to be a feature of most presentations of sentence logic at an elementary level.

Why does the confusion persist? It probably does so because of the way mathematical theorems are stated.

Consider a simple theorem in number theory:

prove that either $n$ is prime or $2^n - 1$ is composite.

Such a statement involves two concealed conventions.

(1) Hidden in the enunciation, there is obviously an implicit universal quantification '$\forall n$' (for all $n \in \mathbf{P}$), which logicians would prefer to see stated explicitly, but which, in contexts other than logic, is often taken for granted in such theorems.

(2) The other is more subtle. If $p$ is written for '$n$ is prime' and $q$ is written for '$2^n - 1$ is composite', what the question *appears* to be asking us to do is to

prove that $p \vee q$.

For general $p$ and $q$, however, '$p \vee q$' is just another proposition and, were $p$ and $q$ independent, '$p \vee q$' could be either false or true. But what the theorem is actually saying is that $p$ and $q$ are *not* independent, that there is a certain relation between them. The challenge really being offered is to

prove that $p \vee q$ *is true*,

that is, that the *assertion* of '$p \vee q$' is a valid deduction from the axioms of the subject, or, in other words, that it is impossible for $p$ and $q$ to be false at the same time. The particular relation between $p$ and $q$ in this case is '$p$ and $q$ are exhaustive'. Thus, borrowing the notation proposed for this relation in Chapter 16, what the question *actually* says is

$$\text{prove that} \quad p \vee\kern-1em\vee q.$$

Being a relation, $\vee\kern-1em\vee$ does not have a truth table (unlike the operation $\vee$). For $u$ real, both the sentences

$$\text{`}u < 7 \quad \vee \quad u > 3\text{'}$$

and

$$\text{`}u < 3 \quad \vee \quad u > 7\text{'}$$

can be written down and their truth values investigated for any $u$, but whereas

$$\text{`}u < 7 \quad \vee\kern-1em\vee \quad u > 3\text{'}$$

is a valid statement,

$$\text{`}u < 3 \quad \vee\kern-1em\vee \quad u > 7\text{'}$$

is not, because the first of the previous pair of sentences is *always true*, whereas the second *can* be false. Similarly, with '$r \Uparrow s$' used for the relation '$r$ and $s$ are mutually exclusive', one can correctly write

$$\text{`}u < 3 \quad \Uparrow \quad u > 7\text{'},$$

but *not*

$$\text{`}u < 7 \quad \Uparrow \quad u > 3\text{'}.$$

The existence of these *two* hidden conventions (1) and (2) means that in many applications to mathematical reasoning '$p \vee\kern-1em\vee q$' is *effectively* saying, not merely that '$p \vee q$ is true', but that '$p \vee q$ is *always* true', and the same thing can be said about '$p \rho q$' for other relations $\rho$. It is, however, important to keep these two assumptions separate in one's mind, because this notion of 'always' is due to assumption (1), the implicit presence of some universal quantifier(s), and is *not* inherent in the relation itself, as can be clearly seen when an existential quantifier is present. For example,

$$\forall r \; \exists s \{ r + s \; \veebar \; r' \}$$

says that

$$\forall r \; \exists s \{ \text{'}(r + s) \vee r''  \text{ is true} \}$$

and here, obviously, '. . . is true' cannot be replaced by '. . . is always true'. The quantified statement is, in fact, true—although this is not germane to the argument; this can be seen by taking $s$ to be $r'$ or, indeed, $s$ and $r$ to be mutually exclusive. It should by now be unnecessary to reiterate that

$$\forall r \; \exists s \{ (r + s) \vee r' \}$$

is as nonsensical a statement as the earlier examples already criticized.

The use of less familiar relations to start with has been deliberate, in order to suppress preconceived expectations. Suppose now, however, that the same number theory result is written, in a more usual form, as:

prove that $2^n - 1$ is prime only if $n$ is prime.

Let $p$ stand, as before, for '$n$ is prime' and let $r$ denote '$2^n - 1$ is prime'; (thus $r$ is $q'$). The question now *appears* to be saying

prove that $\quad r \rightarrow p,$

('$r$ only if $p$'), whereas what it is *actually* saying is

prove that $\quad r \rightarrow p$ *is true*,

that is,

prove that $\quad r \Rightarrow p,$

('$r$ implies $p$'), thereby expressing the fact that $r$ and $p$ are related in a certain way and are not independent. It is a relation of implication ($\Rightarrow$) that is at issue in *all* mathematical theorems stated in this way, *never* a mere conditional connective ($\rightarrow$).

Notice that the sentence '$p \rightarrow r$' is no less respectable an object of study than '$r \rightarrow p$'. But '$p \rightarrow r$' is sometimes true and sometimes false. For example, if $n = 11$ (prime), then $2^n - 1 = 2047$ (composite) and so

$p$ is true, $r$ is false and '$p \rightarrow r$' is false. Hence the converse implication '$p \Rightarrow r$' is *not* valid. But that does not disqualify the compound proposition '$p \rightarrow r$' from being investigated.

**17.9**

In a famous and influential aphorism, Bertrand Russell defined pure mathematics as the class of all propositions of the form '$p$ implies $q$'.

The writer who pretends that '$p$ implies $q$' has a truth table (sic)

| $p$ | F | T | F | T |
|---|---|---|---|---|
| $q$ | F | F | T | T |
| $p$ implies $q$ | T | F | T | T |

is claiming that '$p$ implies $q$' is a well-defined proposition, *whatever* the propositions $p$ and $q$ are, not an assertion that $p$ and $q$ are related in a particular way. It might be possible to rescue the epigram by saying that pure mathematics for such a writer is the class of all propositions of the form '$p$ implies $q$' that are true. But surely the way Russell used 'implies' corresponds exactly to the meaning we *want* it to have. The definition must be framed to secure this, which is *not* what happens when it is claimed that 'implies' has a truth table.

**17.10**

Corresponding to the connective $\leftarrow$ ('if') is another relation $\Leftarrow$ :

$$\text{'}p \Leftarrow q\text{'} \quad \text{means that} \quad \text{'}p \leftarrow q \text{ is true'}.$$

The best reading for '$p \Leftarrow q$' is '$p$ follows from $q$'. (This is shorter and more euphonious than '$p$ is implied by $q$'.)

There is an exactly analogous distinction between the connective $\leftrightarrow$ ('if and only if') and the relation $\Leftrightarrow$ ('implies and follows from'). The operation $\leftrightarrow$ (often called the *biconditional connective*) is defined by the truth table for '$p \leftrightarrow q$', namely

| $p$ | 0 | 1 | 0 | 1 |
|---|---|---|---|---|
| $q$ | 0 | 0 | 1 | 1 |
| $p \leftrightarrow q$ | 1 | 0 | 0 | 1, |

and is obviously commutative. It is easily verified that '$p \leftrightarrow q$' can be expressed in any of the equivalent forms

$$\text{`}(p \leftarrow q) \wedge (p \rightarrow q)\text{'}, \quad \text{`}(p \vee q') \wedge (p' \vee q)\text{'}, \quad \text{`}(p \wedge q) \vee (p' \wedge q')\text{'}.$$

Judicious juggling with these expressions will fairly easily prove that the operation $\leftrightarrow$ is associative, but a simpler method is to introduce the addition operation of the boolean ring. As explained in Chapters 15 and 16, this is the 'exclusive or' operation. The sentence '$p \leftrightarrow q$' is the negation of '$p + q$'. Thus

$$\text{`}p \leftrightarrow q\text{'} \quad \text{is the same as} \quad \text{`}1 + p + q\text{'}$$

and the result becomes obvious, since addition is already known to be associative.

The statement

$$\text{`}p \Leftrightarrow q\text{'} \quad \text{means that} \quad \text{`}p \leftrightarrow q \text{ is true'},$$

that is, that '$p$ and $q$ have the same truth value' or '$p$ is logically equivalent to $q$'. Since it is easily verified that this relation $\Leftrightarrow$ is reflexive, symmetric and transitive, $\Leftrightarrow$ is an equivalence relation in the technical sense and the use of the word 'equivalent' causes no conflict. (Note once more the different properties that have to be considered: it would be *meaningless* to ask if $\leftrightarrow$ is transitive or if $\Leftrightarrow$ is associative.)

In a purely logical context, with every proposition denoted by a single letter, it is occasionally acceptable to denote this relation by '$\equiv$', or even '$=$'. Above, for example, it was very tempting to write informally

$$p \leftrightarrow q = (p \leftarrow q) \wedge (p \rightarrow q) = \text{etc.}$$

and

$$p \leftrightarrow q = 1 + p + q$$

to denote such equivalence, even before the notion had been explicitly introduced. Such temptations will be scrutinized later. But it is obvious that this must *never* be done in general mathematical work, where the sentences will usually be written in full, often with the signs '$\equiv$' and '$=$' already occurring within the structure of these statements and denoting other relations—between numbers, etc. So it must *always* be, for example.

$$5x \equiv 6(7) \Leftrightarrow x \equiv 4(7) \quad \text{and} \quad ab = c \Leftrightarrow a = cb^{-1},$$

*never*[†]

$$5x \equiv 6(7) \equiv x \equiv 4(7) \quad \text{and} \quad ab = c \equiv a = cb^{-1}$$

or

$$5x \equiv 6(7) = x \equiv 4(7) \quad \text{and} \quad ab = c = a = cb^{-1}.$$

Having now established a relation ($\Leftrightarrow$) of equivalence within the system, it is apparent that the relation $\Rightarrow$ is antisymmetric with respect to it: if $p \Rightarrow q$ and $q \Rightarrow p$, then $p \Leftrightarrow q$. So, being also reflexive and transitive, $\Rightarrow$ is a partial order relation.

## 17.11

To sum up, the distinction being urged is that between the left and right sides of Table 17.1.

**Table 17.1·**

| Operation (Connective) | | | | Relation | |
|---|---|---|---|---|---|
| Conditionality | $p$ only if $q$<br>if $p$, then $q$ | $p \rightarrow q$ | $p \Rightarrow q$ | $p$ implies $q$<br>$p \rightarrow q$ is true<br>$p$ is a sufficient condition for $q$ | Implication |
| | $p$ if $q$ | $p \leftarrow q$<br>$q \rightarrow p$ | $p \Leftarrow q$<br>$q \Rightarrow p$ | $p$ follows from $q$<br>$p \leftarrow q$ is true<br>$p$ is a necessary condition for $q$ | |
| Biconditionality | $p$ if and only if $q$ | $p \leftrightarrow q$ | $p \Leftrightarrow q$ | $p$ implies and follows from $q$<br>$p \leftrightarrow q$ is true<br>$p$ is a necessary and sufficient condition for $q$<br>$p$ is equivalent to $q$ | Equivalence |

When the work in sentence logic is matched with that in the other concrete systems for which boolean algebra is the mathematical model—using 'model' in the applied mathematician's, not the logician's, sense—the important correlations, noted in Table 16.3, are those among the 3 equivalence relations (reflexive, symmetric and transitive),

$$\Leftrightarrow \text{ between sentences}$$

$$\Leftrightarrow (\text{or} =) \text{ between switches}$$

$$= \text{ between sets,}$$

[†]Similar misuses ot the sign of equality have already been deprecated in Section 1.3.

and those among the 3 partial order relations (reflexive, antisymmetric and transitive),

$\Rightarrow$ between sentences

$\leq$ between switches

$\subseteq$ between sets.

In switching algebra, where '$a \to b$' and '$a \leftrightarrow b$' are switches and '$a \leq b$' and '$a \Leftrightarrow b$' express relations between the switches $a$ and $b$ (and hence are propositions), everyone can see the difference clearly, as they can also in set algebra, where 'A $\to$ B' and 'A $\leftrightarrow$ B' are sets and 'A $\subseteq$ B' and 'A = B' state relations between the sets A and B (and so are again propositions).

In sentence logic, however, '$p \to q$' and '$p \leftrightarrow q$' are propositions, and '$p \Rightarrow q$' and '$p \Leftrightarrow q$', being relations between $p$ and $q$, are *also* propositions. It is because everything being considered is a proposition that confusion can and does occur. But '$p \to q$' and '$p \Rightarrow q$' are *different* sentences, as are '$p \leftrightarrow q$' and '$p \Leftrightarrow q$'. What is claimed here is that the differences between the pairs of sentences are really just as significant as the more obvious differences in the other two systems. To fail to discriminate is to perplex the student and runs counter to the careful distinction between operations and relations expected in other branches of mathematics.

## 17.12

In what sense can '$p \Rightarrow q$' be false? Of course, when one writes '$p \Rightarrow q$' one is usually making an assertion of a quite definite kind; if '$p \Rightarrow q$' is false, then usually a deductive error has been made. For example, to pick up (for the last time) the threads of that earlier example, one can certainly *write*

$$'u < -5 \quad \Rightarrow \quad u^2 > 40',$$

but this is false because '$u < -5$' *can* be true and '$u^2 > 40$' false at the same time (for example, when $u = -6$). Now contrast this with

$$'u < -5 \quad \to \quad u^2 > 40'$$

and you will appreciate the following fundamental distinction.

The sentence '$p \to q$' is (in general) sometimes true and sometimes false, *depending on the truth values of the individual propositions p and*

$q$; specifically, '$p \rightarrow q$' is false in those circumstances in which $p$ is true and $q$ is false. The sentence '$p \Rightarrow q$', however, is false *if and only if there exist* ANY *circumstances* in which $p$ is true and $q$ is false. (Think carefully about this and then write out for yourself some corresponding observations to show the contrast between '$p \leftrightarrow q$ is false' and '$p \Leftrightarrow q$ is false'.)

This observation will explain a point that may have been worrying you. Starting with one sentence '$p \rightarrow q$', another one '$p \Rightarrow q$' was created, defined as '$p \rightarrow q$ is true'. Could one then introduce a further sentence '$p \Rrightarrow q$', defined as '$p \Rightarrow q$ is true', and so on in an unending hierarchy? Mercifully, no. The sentence '$p \Rightarrow q$ is true' merely says that it is never the case that $p$ is true and $q$ is false simultaneously, which is precisely what '$p \Rightarrow q$' itself says, so that one is saved from any such awesome endless regression.

## 17.13

It seems to the present writer that the feature that is lacking in elementary algebraic accounts of sentence logic is a simple singular relation, $\tau$, which converts a *sentence* $p$ into what may be called an *assertion*, $p\tau$. This assertion is still a sentence: it is still capable of being true or false (but will normally be false only if some deductive error has occurred). All assertions are sentences, but not all sentences are assertions. The relation $\tau$ may be read as 'is true'. It is this singular relation that links '$p \rightarrow q$' to '$p \Rightarrow q$' by the requirement that

$$\text{'}p \Rightarrow q\text{'} \quad \text{has to be the same as} \quad \text{'}(p \rightarrow q)\tau\text{'}.$$

Similarly,

$$\text{'}p \Leftrightarrow q\text{'} \quad \text{has to be the same as} \quad \text{'}(p \leftrightarrow q)\tau\text{'},$$

$$\text{'}p + \!\!\!+ q\text{'} \quad \text{has to be the same as} \quad \text{'}(p + q)\tau\text{'}$$

and so on. All such statements of the form '$p \rho q$', where $\rho$ is any of the relations in Table 16.3, are automatically assertions.

The assertion '$r\tau$' says that 'the sentence $r$ is true' and '$r'\tau$' says that '$r$ is false'. The singular relation $\tau$ has the unique property that

$$\text{'}r\tau\tau\text{'} \quad \text{is the same as} \quad \text{'}r\tau\text{'},$$

so that $\tau^n$ ($n \geq 1$) may be replaced by $\tau$.[†]

---

†Note that the analogue of the relation 'is true' in switching algebra would be 'is closed' and in set algebra would be 'is the universal set'. But it is obviously only for the relation $\tau$ *in logic* that $\tau^2$ even makes sense.

A formal proof consists of a chain of *assertions, not* just a collection of unasserted sentences (sometimes called 'propositional forms') offered merely for examination. The crucial point is that *it is meaningless to say that any asserted sentence has a truth table.*

As one further illustration, consider the structure of a proof by mathematical induction. (This topic is discussed in detail in Chapter 18.) If $P(n)$ denotes a sentence involving a natural number $n$, the two results that have to be established in order to prove by mathematical induction the theorem that

$$\text{'P}(n) \quad \text{is true for all} \quad n \in \mathbf{N'}$$

are

$$(1) \ P(0) \text{ is true;}$$

$$(2) \text{ for all } k \in \mathbf{N}, \quad P(k) \ \Rightarrow \ P(k + 1).$$

Both these steps in the demonstration involve *assertions*; and the inductive transition (2) is (once again) an *implication*.

**17.14**

Professional logicians tend to favour axiomatic treatments rather than the algebraic approaches usual in elementary introductions. What has been written above (algebraically) as '$p\tau$' corresponds *approximately* to what logicians might write (axiomatically) as '$\vdash p$' ('$p$ is provable') or as '$\vDash p$' ('$p$ is valid'), these phrases having certain precise technical meanings. Similarly, '$p \Rightarrow q$' corresponds *roughly* to '$\vdash p \rightarrow q$' [which, because of the deduction theorem, is the same as '$p \vdash q$'] (syntactic implication) or to '$\vDash p \rightarrow q$' ['$p \vDash q$'] (semantic implication). Often, however, logicians seem to feel that by just writing down a proposition they automatically assert it (so that, for example, the mere act of writing '$p \rightarrow q$' converts its *meaning* to '$p \Rightarrow q$'!). This is about as useful an idea as adopting the convention that every time one writes '$t - u$' in algebra one actually *means* '$t > u$'!† Moreover, such a custom leaves one with no possible way of writing down any sentence *without* asserting it. Yet this is what they—and we—regularly need to do when discussing, for example, truth tables. This seems to be the source of much of the ambiguity, which, for this writer at least, spoils many attempts to explain this elegant subject to young students.

---

†The analogy here is quite close, because the introduction of a singulary relation $\pi$ meaning 'is positive' converts '$t - u$' to '$(t - u)\pi$' and '$(t - u)\pi$' is '$t > u$'.

**17.15**

Logic is surely the branch of mathematics, even above all others, in which one must try especially hard to set a good example to pupils by using language and notation carefully and precisely. This leads one to ask whether there are any other ways—apart from distinguishing between connectives and relations—in which the elementary presentation of sentence logic can be radically improved. The conventional exposition seems to flash from textbook to textbook like a chain of bonfires heralding an invasion, each bonfire very much like its predecessor.

The main thing that seems obviously to be needed is some means of discriminating between a proposition and its truth value. When a notation is introduced to achieve this, the whole presentation is dramatically sharpened. At first sight, if you have never considered this before, the alteration may seem unnecessarily ponderous, even pedantic, but you will find that, the more you think about it, the more dazzlingly it illuminates those murky areas in the development: the bonfire is replaced by a searchlight.

The idea is simply to maintain *all the time* a distinction between the actual *sentence*, $p$, and the *valuation* or truth value of $p$, which will be denoted by $v[p]$: $v[p] = 0$ when $p$ is false; $v[p] = 1$ when $p$ is true. So, for example, $p$ might be '$u > 7$', or 'Smith teaches Jones' or '$\triangle ABC$ is isosceles'; $v[p]$ will take the value 0 or 1 according to the falsity or truth of $p$. A truth table, when written in the standard, conventional way, either as

$$p \downarrow q \qquad \begin{array}{c|cc} {}_{p}\diagdown{}^{q} & F & T \\ \hline F & T & F \\ T & F & F \end{array} \qquad \text{or as} \qquad \begin{array}{c|cccc} p & 0 & 1 & 0 & 1 \\ q & 0 & 0 & 1 & 1 \\ \hline p + q & 0 & 1 & 1 & 0 \end{array},$$

is really absolute nonsense. If $p$ is 'Smith teaches Jones', there is no sensible way in which one can say that '$p = F$' or '$p = 0$' or even that 'the variable $p$ is zero'. It is $v[p]$, not $p$ itself, that takes the value 0 or 1.

To confuse $p$ and $v[p]$ is somewhat like writing a matrix as

$$\begin{bmatrix} 11 & 12 & 13 \\ 21 & 22 & 23 \end{bmatrix},$$

using $rs$ rather than $a_{rs}$ for its elements, and thereby making the symbol $rs$ denote both an actual position *and* the numeric value of the element occurring *at* that position—and nobody would advocate doing that, would they?

**17.16**

What a truth table does is to define $v[p \circ q]$, for a particular operation $\circ$, in terms of $v[p]$ and $v[q]$. *All* truth tables, therefore, should have $v[p]$ and $v[q]$ for their headings, *not p and q*, thus:

$v[p \nrightarrow q]$

| $v[p]$ \ $v[q]$ | F | T |
|---|---|---|
| F | F | F |
| T | T | F |

or

| $v[p]$ | 0 | 1 | 0 | 1 |
|---|---|---|---|---|
| $v[q]$ | 0 | 0 | 1 | 1 |
| $v[p \uparrow q]$ | 1 | 1 | 1 | 0 |

This small change of emphasis is all important.

Similarly, if $p$ and $q$ are sentences, to talk about '$f(p, q)$', as we allowed ourselves to do earlier, is really quite unjustified until that curious object has been properly defined. In Table 16.2, $f(x, y)$ is defined when and only when $x, y \in \{0, 1\}$. In its application to logic, what $f(x, y)$ tabulates is simply $f(v[p], v[q])$: '$f(p, q)$' can only be given a meaning by defining it as the sentence whose valuation $v[f(p, q)]$ is $f(v[p], v[q])$.

Thus it is the operations between the valuations that *determine* the propositional connectives. [A similar observation in reference to set operations was made in Section 16.7: their interrelation will be examined presently.] The connective &, for example, between two sentences is introduced by selecting the appropriate one of the 16 possible binary operations—the one denoted by '∧' in Table 16.2—and *defining* '$p$ & $q$' to be the sentence with the property that

$$v[p \ \& \ q] \quad \text{is} \quad v[p] \wedge v[q].$$

(Here, a less familiar sign '&' for conjunction has deliberately been used to drive the point home. After definition, one can blur the distinction by replacing '$p$ & $q$' by '$p \wedge q$'.)

All the propositional connectives are defined in this way: *an operation between the valuations induces an operation between the propositions.*

**17.17**

Further clarification now follows at once as soon as one recognizes that *a relation between the valuations induces a relation between the propositions.*

When, a few pages ago, '$p \vee\hspace{-0.4em}\vee q$' was written to indicate that '$p$ and $q$ are exhaustive', the relation $\vee\hspace{-0.4em}\vee$ between $p$ and $q$ was being defined (although it was glossed over at the time) by the requirement that the valuations $v[p]$ and $v[q]$ should be exhaustive, that is,

$\qquad$ '$p \vee\hspace{-0.4em}\vee q$' $\qquad$ is defined to be the same as $\quad$ '$v[p] \vee\hspace{-0.4em}\vee v[q]$',

where

$\qquad$ '$v[p] \vee\hspace{-0.4em}\vee v[q]$' $\quad$ is defined to be the same as $\quad$ '$v[p] \vee v[q] \equiv 1$'.

Because of the earlier definition, '$v[p] \vee v[q] \equiv 1$' can be replaced by '$v[p \vee q] \equiv 1$' or, using the singulary relation $\tau$ above, by '$(p \vee q)\tau$'. But it cannot be replaced by '$p \vee q \equiv 1$' or by '$p \vee q = 1$' because, now that a sensible nomenclature has been adopted, statements like '$r \equiv 1$' or '$r = 1$' are meaningless for any sentence $r$.

In other words, a particular one of the possible abstract relations (here $\vee\hspace{-0.4em}\vee$) is being selected and, from the property

$$\text{'}x \vee\hspace{-0.4em}\vee y\text{'},$$

(which says that '$x \vee y \equiv 1$'), a corresponding relation

$$\text{'}p \vee\hspace{-0.4em}\vee q\text{'},$$

(also denoted for economy's sake by $\vee\hspace{-0.4em}\vee$), is being *imposed on the sentences* $p$ and $q$, by taking $x$ as $v[p]$ and $y$ as $v[q]$. Try carrying out for yourself a similar exercise to explain '$p \Uparrow q$' ('$p$ and $q$ are mutually exclusive').

The relation of implication between propositions is created when

$\qquad$ '$p \Rightarrow q$' $\qquad$ is defined to be the same as $\quad$ '$v[p] \Rightarrow v[q]$',

where

$\qquad$ '$v[p] \Rightarrow v[q]$' $\quad$ is defined to be the same as $\quad$ '$v[p] \rightarrow v[q] \equiv 1$'.

So '$p \Rightarrow q$' is defined *from* the abstract relation '$x \Rightarrow y$' by again taking $x$ as $v[p]$, $y$ as $v[q]$.

Similarly, the relation of logical equivalence between propositions is introduced when

$\qquad$ '$p \Leftrightarrow q$' $\qquad$ is defined to be the same as $\quad$ '$v[p] \Leftrightarrow v[q]$', where

'$v[p] \Leftrightarrow v[q]$'    is defined to be the same as    '$v[p] \leftrightarrow v[q] \equiv 1$'.

Observe that the relation '$v[p] \leftrightarrow v[q] \equiv 1$' expresses the fact that the valuations of $p$ and of $q$ are necessarily equal and this is being written '$v[p] \Leftrightarrow v[q]$'. It would be misguided to try to write this as '$v[p] = v[q]$', because that statement would merely say that the valuations of $p$ and $q$ happened to be equal in certain circumstances. For example, one could say

$$\text{if } v[p] \neq v[q], \quad \text{then} \quad v[p + q] = v[p \vee q],$$

whereas one could *not* say

$$\text{'}v[p + q] \Leftrightarrow v[p \vee q]\text{'},$$

because obviously '$v[p + q] \leftrightarrow v[p \vee q]$' is *not always* 1.

The urge to denote *every* equivalence relation by '=' is a perennial trap for the unwary student and is the cause of many of the natural shocks that mathematics teaching is heir to.

Even more cogent is the argument for *never* replacing '$p \Leftrightarrow q$' by '$p = q$', even though it does mean that one has to *resist* the temptation to alter precise statements like '$p \to q \Leftrightarrow p' \vee q$' into superficially simpler-looking ones like '$p \to q = p' \vee q$'.


## 17.18

If you have been an exceptionally attentive reader, you will have had the satisfaction of being able to reprove this writer for not following these precepts himself and you are quite right: several liberties were taken in Chapter 16. But it did not seem quite appropriate to interpose this discussion at that stage and it was judged that these considerations, which have a really powerful impact only in logic, could be temporarily ignored. In Table 16.2, for example, there are remarks like '$x \downarrow y = x' \wedge y''$' to indicate informally that the expressions '$x \downarrow y$' and '$x' \wedge y''$' are equivalent and have the same table of values. More correctly, one ought to have written '$x \downarrow y \Leftrightarrow x' \wedge y''$', because the sign '=' might denote a more fortuitous type of equality than the complete equivalence intended there and it is useful not to relinquish the freedom to make such a distinction. For example, one might want to write things like

$$\text{if } x = y, \quad \text{then} \quad x \downarrow y = x \uparrow y,$$

whereas the claim

$$\text{`}x \downarrow y \quad \Leftrightarrow \quad x \uparrow y\text{'}$$

would obviously be illicit. Similarly, in the identities (1) and (2) in Section 16.4, when mentioning the constant functions $f(x, y) = 0$ and 1, in various asides such as $x + y = (x \leftrightarrow y) \vee (x \nleftrightarrow y)$, and even in the definition $x \oplus y = (x \circ y)'$ in Section 16.3, the use of '$\Leftrightarrow$' for '$=$' would have been more accurate. There are many occasions like these when the context makes the meaning clear and one's slipshod mathematics can pass unnoticed. Unfortunately, however, there are also those other occasions when one's whole understanding of a topic depends on perceiving precisely such fine distinctions as that between '$v[p] = v[q]$' and '$v[p] \Leftrightarrow v[q]$', and then, because one has got into the habit of confusing two ideas, one's ability readily to disentangle them when disentanglement is crucial has atrophied.

## 17.19

It is appropriate now to take another look at set algebra. The characteristic functions referred to in Section 16.6, such as

$$I \to \{0, 1\} : v_A,$$

defined by

$$v_A(t) = 1 \quad \text{if} \quad t \in A \quad \text{and} \quad v_A(t) = 0 \quad \text{if} \quad t \notin A,$$

closely resemble the valuations such as $v[p]$ that have just been examined and the similarity can be strengthened if the function $v_A$ is rewritten as $v[A]$.

Suppose a sentence $p$ is of the form $P(t)$ where P is a predicate and $t$ is an element of some set I. Define a function

$$I \to \{0, 1\} : v_p$$

by

$$v_p(t) = 1 \quad \text{if} \quad v[P(t)] = 1 \quad \text{and} \quad v_p(t) = 0 \quad \text{if} \quad v[P(t)] = 0.$$

For example, if $I = \mathbf{R}$ and $p$, or $P(t)$, is '$t^2 - 7t + 10 < 0$', then $v_p(t) = 1$ iff $t \in (2, 5)$.

The closeness of this correspondence explains a remark in Section

16.1 about there being less difference between set algebra and the 2-valued boolean algebras than there appears to be on the surface. If you go further and associate with each such sentence $p$ the subset $S[p]$ of I defined by

$$S[p] = \{t \in I : v_p(t) = 1\},$$

then the equivalence of two sentences, $p \Leftrightarrow q$, where $p$ and $q$ are of the form P($t$) and Q($t$) and in each case $t \in I$, can be expressed in terms of the equality of two sets, $S[p] = S[q]$. This is just the reverse of the remark that if A and B are 2 subsets of a given set I and $s[A]$ denotes the sentence '$t \in A$', then the equality of two sets, A = B, can be replaced by the equivalence of two sentences, $s[A] \Leftrightarrow s[B]$.

More generally, note that if $\rho$ is *any* of the binary boolean relations in Table 16.3, then

$$p \; \rho \; q \quad \text{can be replaced by} \quad S[p] \; \rho \; S[q],$$

$$A \; \rho \; B \quad \text{can be replaced by} \quad s[A] \; \rho \; s[B].$$

Similarly, if o is any of the binary boolean operations in Table 16.1,

$$S[p \circ q] \; = \; S[p] \circ S[q],$$

$$s[A \circ B] \Leftrightarrow s[A] \circ s[B],$$

which ties everything up very nicely.

## 17.20

The misunderstandings about elementary logic that can spring from the failure to maintain a distinction between $p$ and $v[p]$ and the attempt to make one symbol serve two purposes have their counterparts in switching algebra, although, for various reasons, the problems are less obstrusive there. Strictly, however, $a$ should denote an actual *switch*, and the *valuation* or closure value of the switch $a$ should always be denoted by $v[a] : v[a] = 0$ when $a$ is open; $v[a] = 1$ when $a$ is closed. One should *not* say (when constructing closure tables, for example) that $a = 1$, but only that $v[a] = 1$, so you can now go back and correct some of the remarks in Section 16.11. Like truth tables, the headings of closure tables should ideally *always* be $v[a]$ and $v[b]$, *not* $a$ and $b$. The operations on switches (the 'switching connectives', if you like) are defined from the abstract boolean operations by taking $x$ as $v[a]$ and $y$ as $v[b]$ and the possible relations between switches are likewise induced on the switches from the abstract relations between their valuations by the same substitutions (as recorded in the headings of Table 16.3). The details are exactly parallel to those discussed earlier for sentence logic.

One can maintain a useful distinction between equality of closure values and equivalence of switches by representing the relation '$v[a] \Leftrightarrow v[b]$' by '$a \Leftrightarrow b$' and not, as people tend to do in this subject, by '$a = b$'. Although writing '$a = b$' causes less misunderstanding in switching algebra than a similar liberty does in logic, there is nevertheless the same important point of understanding at issue.

Likewise, statements such as '$v[a] \Leftrightarrow v[b]$' and '$v[a] = v[b]$' again need to be distinguished.

Notice that '$v[c] = 1$' denotes that the switch $c$ happens to be closed (but could conceivably be open in other circumstances) whereas '$v[c] \equiv 1$' says that $c$ is always closed. By grossly overworking the symbol '1', this latter property can also be expressed by '$c \Leftrightarrow 1$'. For example, one might allow oneself to write

$$\text{`}(a \twoheadrightarrow b) \uparrow b \quad \Leftrightarrow \quad 1\text{'},$$

because, whatever the closure values $v[a]$ and $v[b]$ are,

$$v[(a \twoheadrightarrow b) \uparrow b] \equiv 1$$

and so the switch on the left is closed in all circumstances. (Compare a tautology in logic.) In contrast, although one can say (for example, when $v[a] = v[b]$), that

$$v[(a \twoheadrightarrow b) \uparrow a] = 1,$$

one can *not* replace '$= 1$' by '$\equiv 1$' or write

$$\text{`}(a \twoheadrightarrow b) \uparrow a \quad \Leftrightarrow \quad 1\text{'},$$

because the switch on the left *can* be open (when $v[a] = 1$, $v[b] = 0$).

If you are really determined to alter '$a \Leftrightarrow b$' to '$a = b$', then, of course, you will replace '$c \Leftrightarrow 1$' by '$c = 1$' and the distinction you will then have to sustain is that between '$c = 1$' and '$v[c] = 1$', where the predicate '$= 1$' now has a quite different significance in each case.


## 17.21

It is interesting to observe how close the required nuance then is to that in elementary algebra between an identity and an equation; for example, between

$$t^2 - (t - 1)(t + 1) = 1$$

and

$$t^2 - \qquad 6t \qquad = 1,$$

where again potential ambiguity is caused by our use of the same sign '=' for these different purposes. Careful teachers often replace the first statement above by

$$t^2 - (t - 1)(t + 1) \equiv 1$$

to make a distinction apparent, but rarely have they the resolution to carry out this policy consistently throughout elementary algebra. One wonders whether our very familiarity with this particular abuse blinds us to difficulties such false economy in notation may create for children. If the discussion above has helped you to sort out any puzzles you personally may have had about the various uses and misuses of the sign '=' in switching algebra and in logic, then you may appreciate the need to be on the lookout for helpful ways of guiding your own pupils through similar, if more elementary, labyrinths.

# Chapter 18

# Mathematical induction

---

## 18.1

Induction is the process of assembling evidence as a basis for making a conjecture: mathematical induction is a method of proof. Even though one is often lazy and misses off the adjective 'mathematical' and just talks about 'proving by induction', you should make sure that your students are quite clear about the distinction.

If a circular pancake is divided by repeatedly making straight cuts across the pancake (without moving any pieces between cuts) and you ask a young pupil to find the maximum number of pieces there can be after $n$ cuts, he will probably start by collecting data for small values of $n$ and constructing a table.

| number of cuts | $n$ | 0 | 1 | 2 | 3 | 4 | 5 | ... |
|---|---|---|---|---|---|---|---|---|
| number of pieces | $p_n$ | 1 | 2 | 4 | 7 | 11 | 16 | ... |

If he has met the triangular numbers, he may, on the basis of this *inductive* evidence, be able to *conjecture* that, for all $n \geqslant 0$,

$$p_n = 1 + \tfrac{1}{2}n(n + 1)$$
$$= \tfrac{1}{2}(n^2 + n + 2).$$

In this extremely trivial case, a method of *proving* the conjecture is suggested by the pattern of the evidence itself. This, of course, is not always, or even usually, the case. The inductive evidence in support of the four-colour map conjecture did not immediately provide a method of proof, although a proof has now been found; and, at the time of writing, the conjecture known as Fermat's Last 'Theorem' remains a conjecture to everyone except perhaps Fermat himself, strong though the inductive evidence supporting it is felt to be. Yet in both graph theory and number theory there are many theorems, superficially quite similar to these conjectures, that *do* have easy proofs by mathematical induction.

If, in the pancake problem, a new cut is to produce as many pieces as possible, then it must meet each of the previous cuts at an interior point of the pancake and no three cuts must concur. In that case, the $n$th cut will increase the total number of pieces by $n$.

The *proof*, by *mathematical* induction, now requires the usual two observations.

(1) Since $p_0$ is known to be 1 and this is the value taken by the formula when $n = 0$, the conjecture is true when $n = 0$.

(2) *If* the conjecture is true when $n = k$,

$$p_k = 1 + \tfrac{1}{2}k(k + 1),$$

in which case, by the above remark,

$$
\begin{aligned}
p_{k+1} &= p_k + (k + 1) \\
&= 1 + \tfrac{1}{2}k(k + 1) + (k + 1) \\
&= 1 + \tfrac{1}{2}(k + 1)(k + 2),
\end{aligned}
$$

and so the conjecture is *then* true when $n = k + 1$.

Hence the conjecture is *proved* for all $n \in \mathbf{N}$† by mathematical induction and so becomes a *theorem*.

Notice, as a matter of technique, the convenience—importance, even—of using $k$ and *not* $n$ in the inductive step (2). By keeping $n$ solely for the parameter in the *general* statement and using some *different* letter when the parameter occurs in the induction hypothesis, one is able to make sensible statements such as 'if the proposition is true when $n = k$, then it is true when $n = k + 1$'. The pupil who tries to 'economize' by overworking the letter $n$ is likely to be unable to stop himself writing nonsense like 'the formula is correct when $n = n + 1$'.‡ Occasionally it may seem rather tedious to have to insist on this, but it is the *only* sensible way of making the logical structure of the argument intelligible to pupils.

Proof by mathematical induction is a most powerful weapon in the mathematician's armoury. Obviously, it is only conjectures involving an

---

†For the definitions of the sets **P, N, Z**, see Section 2.9.

‡Recall the remarks on computer programming in Section 1.3.

*integer* variable that can be targets for attack, but a surprisingly large number of such results will yield to an assault based on some form or other of the principle of mathematical induction. One of the purposes of this chapter is to remind you of the variety that exists in such proofs, by bringing together threads you may only have followed in isolation.

## 18.2

But an important question to ask first is: why are so many students unhappy about mathematical induction? The reason is easily discovered: they were repelled by their first encounter with the method. Usually this happened when they proved the binomial theorem! Time and time again, when one asks students what they know about mathematical induction, they tell you that it is some strange jiggery-pokery that has to be used to prove the binomial theorem—and they never understood it properly, anyway. This, quite literally, is the sole experience that some pupils are given of this exciting and powerful technique. If the teacher did make any preliminary remarks about mathematical induction in general, these were obviously so brief that they were not seen to have any wide significance and were assumed to be merely an introduction to the grand climax of proving the binomial theorem by this curious special stratagem. The reason for students' aversion to this important procedure is a cause as much for sorrow as for impatience.

Observe that not only is a proof of the binomial theorem by mathematical induction monstrously unsuitable as a first (or even as an early) example of that art, it also does a serious disservice to the binomial theorem itself. If the coefficients are thought of (in the *first* instance) in factorial form, as $n!/[r! (n - r)!]$, that theorem appears as a mysterious and complicated technical result. All it really is, of course, is just a very simple enumerative statement about what happens when like terms are collected, after using the distributive and other properties to expand $(a + x)^n$. This is clearly seen when the above coefficient is introduced instead as $\binom{n}{r}$, the number of ways of selecting $r$ objects from $n$ distinguishable objects, and is only *later* replaced by the factorial formula. The *first* approach to the binomial theorem must surely be by way of the expansion of

$$(a + x_1)(a + x_2) \ldots (a + x_n) = a^n + a^{n-1} \sum x_1 + a^{n-2} \sum x_1 x_2 + \ldots$$
$$+ a^{n-r} \sum x_1 x_2 \ldots x_r + \ldots + x_1 x_2 \ldots x_n,$$

where $\sum x_1 x_2 \ldots x_r$ denotes the sum of all the distinct products formed $r$ at a time from $x_1, x_2, \ldots, x_n$. (The $\sum$ sign can be avoided if one really wishes.) Then, putting $x_1 = x_2 = \ldots = x_n = x$, $\Sigma x_1 x_2 \ldots x_r$ becomes $\binom{n}{r} x^r$, giving the binomial theorem

$$(a + x)^n = a^n + \binom{n}{1}a^{n-1}x + \binom{n}{2}a^{n-2}x^2 + \ldots + \binom{n}{r}a^{n-r}x^r + \ldots + x^n$$

in a *natural* way.

Proving this theorem by mathematical induction, with the coefficients in their factorial form, is quite the most difficult application of the method that could have been devised for immature minds. In the first place, *two* integer variables, $n$ and $r$, occur: induction† only takes place over one of them, but the other, over which the summation takes place, is a constant distraction. There is also the complication at the beginning and end with the cases $r = 0$ and $r = n$, involving the definition of $0!$. But, worst of all, there is that preliminary (and rather unattractive) algebra that has to be done to prove that, for $1 \leqslant r \leqslant k$,

$$\binom{k}{r-1} + \binom{k}{r} = \binom{k+1}{r},$$

the special case of Vandermonde's theorem on which this proof depends. When the left side is presented to him *in factorials*, as

$$\frac{k!}{(r-1)!\,(k-r+1)!} + \frac{k!}{r!\,(k-r)!},$$

the pupil is often at a loss how to proceed. Of course, it is obvious to *us* that the thing to do with *all* such expressions is to take outside the g.c.d. of the numerators and the l.c.m. of the denominators to give

$$\frac{k!}{r!\,(k-r+1)!}[r + (k-r+1)] = \frac{(k+1)!}{r!\,(k+1-r)!},$$

which is the same as the right side. But this sort of technical expertise does not come easily to a pupil for whom factorials are still relatively unfamiliar; one often has to extricate him from a maze of unhelpful juggling with the factorials before he is (grudgingly) satisfied that he has proved the lemma.

Alarmed by the realization that the proof of the binomial theorem is

---

†We too shall sometimes allow ourselves the luxury of dropping the adjective 'mathematical' in the rest of this chapter.

going to depend on this hard-won result, the student will have decided that the demonstration is difficult—and he is right—even before he has tried to come to grips with the novel and what should be enjoyable new technique: proof by mathematical induction.

We smugly chuckle when we recall how nineteenth century schoolchildren were turned off Euclid's geometry for ever by the occurrence in the Elements at Proposition 1.5 of the 'pons asinorum', that ludicrously complicated proof of the isosceles triangle theorem, and we marvel that the teachers of that time could have been so blinkered as not to perceive the inappropriateness of such material for their young charges at that stage in their mathematics education. But have we really progressed so far in our pedagogy, if we can select a proof of the binomial theorem for our first exposition of the principle of mathematical induction?

## 18.3

What then would provide more suitable early examples? Well, obviously, results on the summation of finite series are one source. Examples like

$$1.6 + 2.7 + 3.8 + \ldots + n(n + 5) = \tfrac{1}{3}n(n + 1)(n + 8)$$

$$\frac{1}{1.4} + \frac{1}{4.7} + \frac{1}{7.10} + \ldots + \frac{1}{(3n - 2)(3n + 1)} = \frac{n}{3n + 1}$$

$$\frac{1}{3} + \frac{7}{15} + \frac{17}{35} + \ldots + \frac{2n^2 - 1}{4n^2 - 1} = \frac{n^2}{2n + 1}$$

$$1^2 - 2^2 + 3^2 - \ldots + (-1)^{n-1}n^2 = (-1)^{n-1}\tfrac{1}{2}n(n + 1)$$

$$\frac{4}{1.3} - \frac{8}{3.5} + \frac{12}{5.7} - \ldots + (-1)^{n-1}\frac{4n}{(2n - 1)(2n + 1)} = 1 + \frac{(-1)^{n-1}}{2n + 1}$$

$$\ln \tfrac{2}{1} + 2 \ln \tfrac{3}{2} + 3 \ln \tfrac{4}{3} + \ldots + n \ln \frac{n + 1}{n} = \ln \frac{(n + 1)^n}{n!}$$

are all good candidates. Some teachers also like to use mathematical induction to prove the formulae for the sums to $n$ terms of the standard series

$$\Sigma r, \quad \Sigma r(r + 1), \quad \Sigma r(r + 1)(r + 2), \quad \Sigma \frac{1}{r(r + 1)}, \quad \Sigma \frac{1}{r(r + 1)(r + 2)}$$

$$\text{and} \quad \Sigma r^2, \quad \Sigma r^3.$$

But with most of these series it is probably better to wait and obtain the results by direct methods later.

Some formulae for the sums of finite trigonometric series are reasonably easy to prove by induction; for instance (provided $x \neq r\pi$, $r \in \mathbf{Z}$),

$$\cos x + \cos 3x + \cos 5x + \ldots + \cos(2n-1)x = \tfrac{1}{2}\csc x \sin 2nx$$

$$\tan x + 2\tan 2x + 4\tan 4x + \ldots + 2^{n-1}\tan(2^{n-1}x) = \cot x - 2^n\cot(2^n x).$$

And, if you would really like to give pupils some practice with factorials,

$$1.1! + 2.2! + 3.3! + \ldots + n \cdot n! = (n+1)! - 1$$

$$\frac{1}{2!} + \frac{2}{3!} + \frac{3}{4!} + \ldots + \frac{n-1}{n!} = 1 - \frac{1}{n!}$$

$$2.1! + 5.2! + 10.3! + \ldots + (n^2+1) \cdot n! = n \cdot (n+1)!$$

$$\frac{2}{2!} + \frac{7}{3!} + \frac{14}{4!} + \ldots + \frac{n^2-2}{n!} = 3 - \frac{n+2}{n!}$$

are much more straightforward as starters than the binomial theorem.

## 18.4

The value of these examples is immeasurably increased if, for some of them, the pupil is *not told* the formula for the sum, but is expected to *conjecture* it himself from evidence he collects for small values of $n$. This puts him into the position of a genuine mathematician, who works in this way himself, making a conjecture and often, when this involves an integer variable, testing it by mathematical induction.

Such activity, of course, must be carefully prepared and monitored by the teacher, since only relatively few series have a formula for the sum to $n$ terms and fewer still have expressions that are simple enough for the student to guess without an unreasonable amount of hard work. But several of the examples above will suggest promising material for this treatment and these offer better prospects for enlightenment than sets of questions where *all* the answers are given him in advance. The intelligent pupil is not fooled; he knows that you have not guessed these formulae at all, but have discovered them by some secret process (in fact, by using a method of differences), whose mysteries are not yet being revealed to him.

**18.5**

As a change from series, various formulae involving finite products can easily be established by induction; for instance,

$$\left(1 - \frac{1}{2^2}\right)\left(1 - \frac{1}{3^2}\right)\left(1 - \frac{1}{4^2}\right) \cdots \left(1 - \frac{1}{n^2}\right) = \frac{n+1}{2n}$$

$$\left(1 + \frac{1}{2}\right)\left(1 + \frac{1}{4}\right)\left(1 + \frac{1}{16}\right) \cdots \left(1 + \frac{1}{2^{2^{n-1}}}\right) = 2\left(1 - \frac{1}{2^{2^n}}\right)$$

$$(a + b)(a^2 + b^2)(a^4 + b^4) \cdots (a^{2^n} + b^{2^n}) = \frac{a^{2^{n+1}} - b^{2^{n+1}}}{a - b} \qquad (a \neq b)$$

$$\cos x \cdot \cos 2x \cdot \cos 4x \quad \cdots \quad \cos(2^{n-1}x) = \frac{1}{2^n}\sin(2^n x)\csc x$$

$$(x \neq r\pi, \quad r \in \mathbf{Z}).$$

Inequalities, however, also provide a fruitful area of application, which may easily be overlooked; for example,

$$(1 + x)^n \geq 1 + nx \qquad \qquad \text{for} \quad x > -1, \quad n \geq 0$$

$$\frac{1}{n+1} \leq \frac{1 \cdot 3 \cdot 5 \ldots (2n-1)}{2 \cdot 4 \cdot 6 \ldots (2n)} < \frac{1}{\sqrt{(2n+1)}} \qquad \text{for} \quad n \geq 1$$

$$(a + b)^n \leq 2^{n-1}(a^n + b^n) \qquad \text{for} \quad a > 0, \quad b > 0, \quad n \geq 1$$

$$2\sqrt{(n+1)} - 2 < 1 + \frac{1}{\sqrt{2}} + \frac{1}{\sqrt{3}} + \ldots + \frac{1}{\sqrt{n}} \leq 2\sqrt{n} - 1 \qquad \text{for} \quad n \geq 1.$$

When the students are a little more advanced and know the standard logarithm inequality

$$\frac{x}{1+x} < \ln(1 + x) < x \qquad \qquad (\text{for} \quad x > -1, \quad x \neq 0),$$

they can then use induction to obtain the useful result

$$\ln(n + 1) < 1 + \frac{1}{2} + \frac{1}{3} + \ldots + \frac{1}{n} \leq 1 + \ln n \qquad \qquad \text{for} \quad n \geq 1.$$

[The left part of this inequality yields the immediate corollary that

$\Sigma\ 1/n$ diverges and offers a possible alternative to the usual proof, which involves showing that $s_{2^m} \geq 1 + \frac{1}{2}m$.]

Results that do not 'start' with $n = 0$ or $n = 1$ can be particularly useful for clarifying the logical structure of an induction proof, because they emphasize that it is not merely the inductive transition, $P(k) \Rightarrow P(k + 1)$, that has to be established; for instance,

$$n! > 3^n \qquad \text{for} \quad n \geq 7$$

$$\binom{2n}{n} < 4^{n-1} \qquad \text{for} \quad n \geq 5$$

and so on.

### 18.6

There are many results on divisibility that have interesting proofs by mathematical induction and you may find these useful as examples to use with some classes.

$$8 \mid 9^n + 7 \qquad\qquad (n \in \mathbf{N})$$

$$9 \mid 4^n - 3n - 1 \qquad\qquad (n \in \mathbf{N})$$

$$7 \mid 3^{2n-1} + 2^{n+1} \qquad\qquad (n \in \mathbf{P})$$

$$25 \mid (5n + 2)11^n - 2 \qquad\qquad (n \in \mathbf{N})$$

$$49 \mid 2^{3n+2} + 21n - 4 \qquad\qquad (n \in \mathbf{N})$$

$$5 \mid 3^{2n+1} - 2^{2n-1} \qquad\qquad (n \in \mathbf{P}).$$

These properties can, of course, be conjectured by the pupil if he is given the expressions involving $n$; but all such examples have the obvious disadvantage that the 'secret' of how suitable formulae are discovered is being withheld.

Before starting, the solver may need to be reminded that, if $m \in \mathbf{N}$, then

$$\text{for all} \quad m, \qquad a - b \mid a^m - b^m,$$

$$\text{for odd} \quad m, \qquad a + b \mid a^m + b^m,$$

$$\text{for even} \quad m, \qquad a + b \mid a^m - b^m,$$

these being the consequences of algebraic, rather than arithmetic, identities.

For example, suppose it is required to prove that

$$64 \mid 9^n - 8n - 1 \quad \text{for} \quad n \in \mathbf{N}.$$

It is convenient to let

$$f(n) = 9^n - 8n - 1.$$

The pupils' first observation will be that

$$64 \mid f(0).$$

Now

$$f(k + 1) - f(k) = 8.9^k - 8$$
$$= 8(9^k - 1)$$

and so, using the first lemma above,

$$64 \mid f(k + 1) - f(k) \quad \text{for all} \quad k \geqslant 0.$$

Thus, for $k \geqslant 0$,

$$64 \mid f(k) \quad \Rightarrow \quad 64 \mid f(k + 1),$$

and hence the result

$$64 \mid f(n)$$

is established for all $n \in \mathbf{N}$ by mathematical induction.

Notice particularly that it is by no means necessary to take the combination $f(k + 1) - f(k)$ in order to establish the inductive step. Some other expression

$$\lambda f(k) + \mu f(k + 1)$$

often gives the result more efficiently. For instance, in the worked example above,

$$f(k + 1) - 9f(k) = 64k$$

can be used instead; and, in the last example in the above list, with

$$g(n) = 3^{2n+1} - 2^{2n-1},$$

any of the combinations

$$g(k) + g(k + 1), \quad g(k + 1) - 4g(k), \quad g(k + 1) - 9g(k)$$

works better than $g(k + 1) - g(k)$.

Some (more advanced) examples of this behaviour, however, can easily put too high a premium on ingenuity and divert one from the original aim of providing simple illustrative examples of the method of proof by mathematical induction.


## 18.7

You must look out for the pupil who mixes up his logic by saying

$$64 \mid f(k + 1) \quad \Rightarrow \quad 64 \mid f(k),$$

which, even though it happens to be true, is entirely irrelevant to the induction argument. If this sort of mistake goes undetected, the perpetrator may soon graduate to 'proving' that $n! \leq 2^n$, by saying (1) that this statement is correct when $n = 1$ (which is true), and (2) that

$$(k + 1)! \leq 2^{k+1} \quad \Rightarrow \quad k! \leq \frac{2^{k+1}}{k + 1} \leq 2^k \quad \text{whenever} \quad k \geq 1,$$

(which is also true). But this reasoning, of course, proves nothing useful and, in fact, the conjecture itself is true only when $0 \leq n \leq 3$ and is false for *all* $n \geq 4$. (This sort of fallacious 'proof' is a useful example to keep up your sleeve to convince any students that fall into this trap of the error of their ways.)


## 18.8

That nice little result about the sum of the cubes of three consecutive integers always being a multiple of 9 is particularly interesting because it is slightly more easily proved by induction than it is directly; do try both methods for yourself. Again, one simple way of showing that, if

$$t_n = \frac{1}{n + 1} + \frac{1}{n + 2} + \ldots + \frac{1}{2n},$$

then

$$t_n = 1 - \frac{1}{2} + \frac{1}{3} - \frac{1}{4} + \ldots + \frac{1}{2n - 1} - \frac{1}{2n}$$

is to use mathematical induction (although here a direct proof is equally straightforward). There really is no shortage of excellent material!

The best examples for proof by induction are undoubtedly those where it is easy to make the conjecture but more difficult to find a direct proof (that is, a proof that does not depend on anticipating the possible result). Recurrence relations are a valuable source of such questions: if carefully chosen they can provide very good examples for successful conjecture. Here are two simple problems involving only first order recurrence relations. [See Section 1.4 for a note on this terminology.]

The sequences $(u_n)$ and $(v_n)$ are defined by

$$u_0 = \tfrac{2}{3}, \qquad u_{n+1} = \frac{1}{2}\left(3 - \frac{1}{u_n}\right) \qquad (n \geqslant 0);$$

$$v_0 = 2, \qquad v_{n+1} = (n + 1)v_n - n \qquad (n \geqslant 0).$$

Guess the formulae for $u_n$ and for $v_n$ and prove your conjectures correct by mathematical induction.

$$\left[ u_n = \frac{2^n + 1}{2^n + 2} \; ; \qquad v_n = n! + 1. \right]$$

The sequence of Fibonacci numbers

$$0, 1, 1, 2, 3, 5, 8, 13, 21, 34, 55, 89, 144, \ldots,$$

which is defined by a second order recurrence relation,

$$f_0 = 0, \quad f_1 = 1, \quad f_{n+1} = f_n + f_{n-1},$$

can be made to yield a rich harvest.

The simple but important result that consecutive Fibonacci numbers are always coprime is proved by induction, by observing that

(1) $f_0$ and $f_1$ are coprime, [or $f_1$ and $f_2$, if you prefer to avoid the zero member], and

(2) if $f_{k-1}$ and $f_k$ are coprime, then $f_k$ and $f_{k+1}$ are coprime.

Examination of the pattern of the sequence should quickly lead to conjectures like

$$2 \mid f_{3r}, \quad 3 \mid f_{4s}, \quad 5 \mid f_{5t}, \quad (r, s, t \in \mathbf{N}).$$

These also can be proved by induction, most easily after establishing (respectively) that

$$f_{n+3} = f_n + 2f_{n+1},$$
$$f_{n+4} = 2f_n + 3f_{n+1},$$
$$f_{n+5} = 3f_n + 5f_{n+1}.$$

These properties are, of course, only the simplest in a chain of similar results of the type

$$f_l \mid f_{lm}.$$

Theorems like

$$f_1 + f_3 + f_5 + \ldots + f_{2n-1} = f_{2n}$$
$$f_2 + f_4 + f_6 + \ldots + f_{2n} = f_{2n+1} - 1$$
$$f_{n-1}f_{n+1} = f_n^2 + (-1)^n$$

and scores of others are obvious candidates for treatment by induction.

The general formula for $f_n$, however, is

$$f_n = \frac{\tau^n - (-\tau)^{-n}}{\sqrt{5}},$$

where $\tau = \frac{1}{2}(\sqrt{5} + 1)$ is the golden ratio, and this is rather complicated to use as an early example (although it is an excellent choice for later work).

There are instead plenty of other second order recurrence relations for which (provided the initial values are chosen carefully) the general member of the sequence can be readily conjectured by the pupils after working out a few instances, and then the boys and girls can have the satisfaction of proving their conjectures correct by mathematical induction. For example,

$$a_0 = 4, \quad a_1 = 8, \quad a_{n+1} = 6a_n - 5a_{n-1}, \qquad [a_n = 3 + 5^n]$$
$$b_0 = 2, \quad b_1 = 6, \quad b_{n+1} = 6b_n - 8b_{n-1}, \qquad [b_n = 2^n + 4^n]$$

$$c_0 = 3, \quad c_1 = 1, \quad c_{n+1} = 2c_n + 3c_{n-1}, \qquad\qquad [c_n = 3^n + 2(-1)^n]$$

$$d_0 = 1, \quad d_1 = 2, \quad d_{n+1} = (n+2)d_n - (n+1)d_{n-1}, \quad [d_n = \sum_{r=0}^{n} r!].$$

Harder examples can, of course, be used if the students are given the formula to prove, although that rather spoils the fun, except perhaps for $(f_n)$ where the sequence is so fascinating and the formula for $f_n$ unexpectedly complicated.

The interest of these particular problems lies in the need to specify 2 initial values—or $r$ initial values if the recurrence relation is of order $r$–and the consequent modification of the logical structure of the usual induction proof. It becomes the *conjunction* of the propositions P($k$) *and* P($k - 1$) that implies P($k + 1$), and there are *two* initial cases, P(0) and P(1), that have to be checked independently (with, of course, the appropriate extensions when the recurrence relation is of higher order).

### 18.9

On many occasions when induction arguments arise in advanced mathematics, the truth of P($k + 1$) is established not by assuming just the truth of P($k$), but by assuming the truth of P($j$) for *all* $j \le k$; that is, by assuming

$$P(k) \wedge P(k - 1) \wedge \ldots \wedge P(1) \wedge P(0).$$

This type of induction argument can also be illustrated by a recurrence relation; for instance,

$$u_{n+1} = u_0 + u_1 + u_2 + \ldots + u_n, \quad u_0 = 3.$$

The formula $u_n = 3.2^{n-1}$ ($n \ge 1$) is easily conjectured and proved correct. Notice that only *one* 'starting' value is required.

### 18.10

The formula for the derivative of a power of $x$,

$$\frac{\mathrm{d}}{\mathrm{d}x}(x^n) = nx^{n-1} \qquad (n \in \mathbf{Z}) \tag{1}$$

is perhaps the most important formula in sixth form mathematics. Have you noticed what an outstanding contender this result is for proof by mathematical induction? There may, of course, quite often be good reasons for preferring to obtain it from first principles as

$$\lim_{h \to 0} \frac{(x + h)^n - x^n}{h}$$

and, if $n \in \mathbf{P}$, expanding $(x + h)^n$ by the binomial theorem. On the other hand, with some classes one may want to establish (1) at a stage before one has taught the binomial theorem.

It is also arguable that mathematical induction offers anyway the simplest and most efficient proof of (1). After all, one is sure to have started this work by considering piecemeal the separate cases $n = 0$, $n = 1$, $n = 2$, $n = 3$ and perhaps $n = 4$. Thus the formula (1) is almost certain to present itself first as a *plausible conjecture* for $n \in \mathbf{N}$, arising directly out of this accumulated experience. This makes (1) an obvious candidate for a proof by mathematical induction; in fact, it absolutely cries out for such treatment! [One is supposing, of course, that the pupils will have had previous experience of the method as a general procedure: one does *not* want to stop mathematical induction being regarded as an eccentric device for proving the binomial theorem, only to have it relegated to the status of a curious dodge for establishing the formula for the derivative of a power!]

As a preliminary to proving (1) by induction, the result for $n = 1$, that is,

$$\frac{\mathrm{d}}{\mathrm{d}x}(x) = 1, \tag{2}$$

must be obtained from first principles; also, the product rule for derivation,

$$\frac{\mathrm{d}}{\mathrm{d}x}(uv) = v\frac{\mathrm{d}u}{\mathrm{d}x} + u\frac{\mathrm{d}v}{\mathrm{d}x},$$

must be familiar.

Then, writing $x^{k+1}$ as $x^k . x$ and using (2),

$$\frac{\mathrm{d}}{\mathrm{d}x}(x^{k+1}) = x\frac{\mathrm{d}}{\mathrm{d}x}(x^k) + x^k. \tag{3}$$

The induction hypothesis is that

$$\frac{\mathrm{d}}{\mathrm{d}x}(x^k) = kx^{k-1} \qquad (k \geq 1).$$

Then, from (3),

$$\frac{d}{dx}(x^{k+1}) = x \cdot kx^{k-1} + x^k$$

$$= (k+1)x^k.$$

Thus the formula (1) is correct when $n = 1$, and, if it is correct for $n = k$, then it is also correct for $n = k + 1$, so that the formula (1) is proved for all $n \geqslant 1$ by mathematical induction.

Replacing $k$ in (3) by $j - 1$ and rearranging,

$$\frac{d}{dx}(x^{j-1}) = \frac{1}{x}\frac{d}{dx}(x^j) - x^{j-2} \qquad (x \neq 0). \tag{4}$$

Now take a second induction hypothesis: that

$$\frac{d}{dx}(x^j) = jx^{j-1} \qquad (j \leqslant 1).$$

Then, from (4),

$$\frac{d}{dx}(x^{j-1}) = \frac{1}{x} \cdot jx^{j-1} - x^{j-2}$$

$$= (j-1)x^{j-2} \qquad (x \neq 0).$$

Hence, if the formula (1) is correct for $n = j$, then it is correct also for $n = j - 1$. So the formula (1) is proved also for $n < 1$ $(x \neq 0)$ and hence for all $n \in \mathbf{Z}$.

Although two induction arguments have been used *here* to emphasize the full potential of the method, understanding the first part of the proof (for $n \geqslant 1$) would obviously be a sufficient objective for the majority of classes. In any case, most teachers would prefer to get the result for $n \leqslant -1$ either from

$$\frac{d}{dx}\left(\frac{1}{y}\right) = -\frac{1}{y^2}\frac{dy}{dx}$$

or from

$$1 = \frac{d}{dx}(x) = \frac{d}{dx}(x^{m+1} \cdot x^{-m}) = (m+1)x^m \cdot x^{-m} + x^{m+1}\frac{d}{dx}(x^{-m}),$$

rather than by using a second induction. But it is always interesting to explore just what can be done.

### 18.11

There are many proofs in elementary mathematics where an induction argument lies concealed, often behind a phrase such as '. . . and so on'. Take, for example, one of the standard school proofs that the sum of the interior angles of a (convex) $n$-sided polygon is $(n - 2)\pi$. When analysed, this amounts to saying that, given any convex polygon of $(k + 1)$ sides, $A_1A_2 \ldots A_kA_{k+1}$, cutting off the triangle $A_kA_{k+1}A_1$ leaves a convex $k$-sided polygon $A_1A_2 \ldots A_k$. *If* the sum of the interior angles of this latter polygon is $(k - 2)\pi$, *then* that of the larger polygon is $(k - 2)\pi + \pi$. Since the sum of the angles of a 3-sided polygon is $\pi$, the theorem is established for all $n \geqslant 3$ *by mathematical induction*. [The result is, of course, still true if the polygon is not convex; it is just that school proofs tend to avoid the consequent complications. You, however, might find it interesting to consider how such an omission can be rectified.]

Again, in proving de Moivre's theorem for a positive integer index, the result

$$\text{cis } \alpha \, . \, \text{cis } \beta = \text{cis } (\alpha + \beta)$$

is used to produce successively

$$(\text{cis } \theta)^2 = \text{cis } 2\theta, \quad (\text{cis } \theta)^3 = \text{cis } 3\theta, \quad \text{'and so on'.}$$

Before writing down

$$(\text{cis } \theta)^n = \text{cis } n\theta \quad \text{for all} \quad n \in \mathbf{P},$$

you should at the very least *point out* that an induction argument based on

$$\text{cis } k\theta \, . \, \text{cis } \theta = \text{cis } (k + 1)\theta$$

has to be supplied: if it is omitted, that is only done because it is an obvious step, *not* because it is an unnecessary step.

Similarly if, when establishing the Maclaurin expansion of, for example, $f(x) = \ln(1 + x)$, a lecturer works out $f'(x)$, $f''(x)$, $f^{(3)}(x)$ and $f^{(4)}(x)$, and then adds

$$\text{'and hence} \quad f^{(n)}(x) = \frac{(-1)^{n-1}(n - 1)!}{(1 + x)^n} \quad \text{for all} \quad n \geqslant 1\text{',}$$

he is negligent if he fails to observe that the easy induction argument needed to *prove* this conjecture is being deliberately suppressed.


**18.12**

To illustrate the importance of never arguing from particular cases to general theorems, students should be encouraged to reflect on 'results' that happen to be true for certain small values of $n$ but are false in general; for example,

| | |
|---|---|
| $5 \mid 6^n + 4n^4$. | True $n = 1, 2, 3, 4$; false $n = 5$. |
| $2^n \leqslant n^3 + 1$. | True $0 \leqslant n \leqslant 9$; false $n \geqslant 10$. |
| $1 + \dfrac{1}{3} + \dfrac{1}{5} + \ldots + \dfrac{1}{2n - 1} < 2$. | True $1 \leqslant n \leqslant 7$; false $n \geqslant 8$. |
| $2^{2^n} + 1$ is prime. | True $0 \leqslant n \leqslant 4$; false $n = 5$. |
| $n^2 + n + 41$ is prime. | True $-40 \leqslant n \leqslant 39$; false $n = 40, 41$. |
| If $p$ is an odd prime, then the Fibonacci number $f_p$ is prime. | True $p = 3, 5, 7, 11, 13, 17$;  false $p = 19$ [$4181 = 37.113$]. |

Number theory is an obvious and productive hunting ground for speculations of this type.


**18.13**

Some teachers claim to find the logical structure of proofs by mathematical induction easier to explain to pupils if they use a form of words different from the one used so far in this chapter, which, just for convenience, will be referred to as the 'standard' version. This is certainly an area where there is scope for more educational experiment. Possible alternatives may be based either on the so-called *'well-ordering' principle* or on a *'first failure' principle*.

The *well-ordering principle* describes a characteristic property of the set **N** of natural numbers and says:

*every non-empty subset of* **N** *has a least element.*

[This property is, of course, also enjoyed by the set **P** of positive integers, but *not* by the set **Z** of integers.]

The fundamental observation that a strictly decreasing sequence of natural numbers must be finite is exploited in many proofs in mathematics, one of which will be mentioned presently. You will observe that this property can be considered a simple corollary of the well-ordering principle. For, suppose that $u_1 > u_2 > u_3 > \ldots \geqslant 0$. Then, since the set $\{u_1, u_2, u_3, \ldots\}$ must have a least element, $u_s$ say, it follows that all members $u_n$ of the sequence must satisfy $u_1 \geqslant u_n \geqslant u_s$, which implies that the sequence is finite, since all its members are distinct.

By taking for the subset mentioned in the well-ordering principle the set of all natural numbers for which a given proposition is *false*, that principle becomes a *first failure principle:*

*if there is at least one natural number for which a proposition is false, then there is a first natural number for which it is false.*

An argument based on this principle concentrates on that *first* natural number, $m$ say, for which the proposition is assumed false and either (1) creates a contradiction by producing a smaller natural number for which the proposition is also false or (2) shows that the assumption that the proposition is true when $n = m - 1$ (or, sometimes, true for $n = l$ for *all* $l < m$) but false when $n = m$ is itself contradictory.

But this is not the whole story, because one *still* has to check that the proposition is true when $n = 0$. Why is this necessary? Please think carefully about this; it is very important. The point is that, without showing this, one leaves open the possibility that the proposition may be *false* for *all* $n$. The crucial fact is that for every natural number there is at least one smaller natural number *except* for 0, which is unique in having *no* predecessor. This is why the status of the conjecture for $n = 0$ must be examined independently. [One possible snag about first failure arguments from a teaching point of view is that this step tends to be more often forgotten than the corresponding step in the 'standard' version.] When you are using **P** instead of **N**, it is, of course, 1 that has no predecessor; if you are trying to prove a proposition true for $n \geqslant n_0$, then other appropriate modifications to the argument have to be made.

### 18.14

In case you are unsure what a first failure argument looks like when applied to a really elementary problem, consider in detail the explanation that would be used to establish the formula for the pancake sequence $(p_n)$.

Suppose the formula

$$p_n = 1 + \tfrac{1}{2}n(n + 1)$$

is *not* correct for all $n$. Let $m$ be the *first* natural number for which it fails.

Since $p_0 = 1$, the formula is correct for $n = 0$. So $m \neq 0$ and therefore $m - 1$ is a natural number. Since $m$, by hypothesis, provides the *first* failure, the formula is correct when $n = m - 1$, so that

$$p_{m-1} = 1 + \tfrac{1}{2}(m - 1)m.$$

But, as observed at the beginning of the chapter,

$$p_m = p_{m-1} + m,$$

and so

$$p_m = 1 + \tfrac{1}{2}m(m + 1).$$

This contradicts the assumption that the formula is false when $n = m$. So there is no natural number for which the formula is incorrect.

Alternatively, after showing that $m \neq 0$, so that $m - 1$ is a natural number, the proof can be finished off in the following way. Although this version is perhaps not very stylish for the present problem, there are plenty of theorems where it provides a really convincing dénouement.

We know that

$$p_{m-1} = p_m - m,$$

and so if, by hypothesis,

$$p_m \neq 1 + \tfrac{1}{2}m(m + 1),$$

then

$$p_{m-1} \neq 1 + \tfrac{1}{2}(m - 1)m,$$

which means that the formula fails for $n = m - 1$, contradicting the definition of $m$ as the natural number that provides the *first* failure.

It is, of course, essential that the young mathematician does not confuse this perfectly valid reasoning with a fallacious '$P(k + 1) \Rightarrow P(k)$' type of argument when attempting a proof in the standard form. The unsoundness of that pseudo-logic was explained earlier (Section 18.7).


## 18.15

When teaching mathematical induction, one has to decide whether one is going (eventually) to go all out to exploit the rich variety of potential arguments that come within its compass, and so prepare the pupil for most contingencies he is likely to meet later at college and university (the hope being that the very breadth of the experience he acquires in marshalling a variety of arguments will strengthen his understanding of *all* the procedures); or whether one is going to stick rigidly to some precise and hackneyed form of words—presumably that associated with the 'standard' proof, which will certainly help to give him confidence in the method, but may also discourage flexibility and initiative and make it difficult for him to follow basically similar reasoning if it is not presented to him in exactly the form with which he is familiar. It is not an easy decision. For the weaker student, of course, a really firm grasp of just one version will certainly be a satisfactory objective.

### 18.16

The well-ordering and first failure principles are widely used in advanced mathematics and lecturers are prone to grumble if their students do not understand them instantly. This is an unfair complaint, as it takes no account of the extreme subtlety of all induction arguments and the time needed for a student to assimilate the nuances in an unfamiliar version.

These 'non-standard' proofs tend to be most persuasive when the hypothesis being tested is couched in language that says so-and-so is false or impossible or does *not* have such-and-such a property, rather than in language that expresses a more 'positive' conjecture. (This informal remark is admittedly silly: proving that 'P($n$) is true' can *always* be rephrased as 'the negation of P($n$) is false'. But you will doubtless recognize the sort of statement we are trying to categorize. It is difficult to be more precise as it is anyway only a subjective judgement that is being made.)

Good candidates for elementary applications of these methods can be found by looking again at the Fibonacci sequence. One notices, for example, that every third member (starting from $f_0$) is even and that every sixth member is divisible by 8. (Both these facts are easily proved by the induction arguments mentioned earlier.) But there seem to be no members that are divisible by 4 without being divisible also by 8, and so one is led to conjecture that

$$f_n \equiv 4 \ (8) \text{ is impossible.}$$

Testing even a thousand members without finding an occurrence of this phenomenon would not prove that it can *never* happen, however much that evidence might boost one's confidence in the conjecture. So how can it be demonstrated? Before reading the proof below, you may like to try devising one of your own.

From

$$f_{n-2} = f_n - f_{n-1}$$

and

$$f_{n-3} = -f_n + 2f_{n-1},$$

corresponding formulae for $f_{n-4}$, $f_{n-5}$, ... are easily obtained and thus one discovers that

$$f_{n-6} = 5f_n - 8f_{n-1}.$$

If such members of the sequence *do* exist, let $f_m$ be the *first* one for which

$$f_m \equiv 4 \ (8).$$

Then

$$5f_m \equiv 4 \ (8),$$

and so

$$f_{m-6} \equiv 4 \ (8).$$

[The ideas latent in this argument can easily be expressed without congruence notation, if you prefer.] So, if $f_m$ has the postulated property, then so also does $f_{m-6}$; so, therefore, does $f_{m-12}$, and so on. [Note. The proof is *not yet* complete. Why? What remains to be shown?]

Since, however, none of $f_0$, $f_1$, $f_2$, $f_3$, $f_4$, $f_5$, is congruent to 4 modulo 8, it follows that *no* $f_n$ can have that property.

In the same way, it can be proved that $f_n \equiv 6 \ (8)$ is likewise impossible and hence all the remaining even Fibonacci numbers are congruent to 2 modulo 8. There are similar results with other moduli.

### 18.17

The following proof is a good elementary example of a particular style of induction argument, known as the *method of descent*. It demonstrates one of the ways of proving that $\sqrt{2}$, or any similar real number, is irrational.

Suppose, if possible, that

$$\sqrt{2} = \frac{a_1}{b_1},$$

where $a_1$ and $b_1$ are positive integers. Since $1^2 < 2 < 2^2$,

$$1 < \frac{a_1}{b_1} < 2,$$

that is,

$$b_1 < a_1 < 2b_1. \tag{5}$$

Now

$$\sqrt{2} = \frac{a_1}{b_1} = \frac{2b_1}{a_1} = \frac{2b_1 - a_1}{a_1 - b_1} = \frac{a_2}{b_2} \quad \text{(say)},$$

where $a_2 = 2b_1 - a_1$, $b_2 = a_1 - b_1$, both $a_2$ and $b_2$ being integers. Since, using (5),

$$0 < a_2 < a_1, \quad 0 < b_2 < b_1,$$

$a_2$ and $b_2$ are *positive* integers, *less* than $a_1$ and $b_1$ respectively, and such that

$$\sqrt{2} = \frac{a_2}{b_2}.$$

Since this process can be repeated indefinitely, whereas no unending sequence of integers satisfying

$$a_1 > a_2 > a_3 > \ldots \geq 1$$

is possible, the original hypothesis that $\sqrt{2}$ is rational must be false.

Observe, as a matter of interest, that this particular proof of the irrationality of $\sqrt{2}$ does *not* require the assumption that $a_1$ and $b_1$ are chosen to be coprime.

The argument in this proof can be cast in an alternative (but equivalent) mould, making it more directly dependent on the well-ordering principle. A pupil may (or may not) find this version of the demonstration more convincing. Instead of taking (as above) $a_1, b_1$ to be *any* integers whose supposed quotient is $\sqrt{2}$, one starts by observing that, if there do exist positive integers $a, b$ such that

$$\sqrt{2} = \frac{a}{b},$$

then the set of possible numerators, $a$, is non-empty and so, by the well-ordering principle, this set must contain a least element, $a_1$ say. It is *this* $a_1$ that is taken as the basis for the above reasoning. Then, when $a_2$, $b_2$ have been defined as before and it turns out that $0 < a_2 < a_1$, a contradiction has been produced. Hence the set of possible numerators is empty and $\sqrt{2}$ is irrational.

This method of descent is associated particularly with the name of Pierre de Fermat (1601–1665), who introduced it into number theory and with it achieved some spectacular successes in proving important theorems. For example, the equation

$$x^4 + y^4 = z^2$$

has no solution in integers (none of which is zero). This is proved by supposing the equation to be soluble and letting $z_1$ be the least positive integer value of $z$ for which a solution exists. From $z_1$, another integer $z_2$ is constructed for which the equation is also soluble. But the procedure chosen makes $0 < z_2 < z_1$, so that a contradiction is obtained in the same way as above, and the equation is thereby proved to be insoluble in integers. The details of the proof are very elegant; you will find them in almost any book on number theory.

### 18.18

One of the aims of this chapter has been to stress the importance of students distinguishing between (1) results they have conjectured on inductive evidence and *believe* to be true and (2) results they have *proved* by *mathematical* induction. So, to conclude this chapter, here is an amusing little example, which you may not know, to illustrate the dangers of overhasty conjecture. It seems to have originated as a problem in a Russian Mathematics Olympiad.

Having discovered a sequence whose first members are

$$1, 2, 4, 8, 16, \ldots,$$

the urge to continue it 32, 64, . . . is well-nigh irresistible. One realizes, of course, that there is, in principle, no bound to the number of possible sequences that start in that way; for instance

$$2^{n-1} + (n-1)(n-2)(n-3)(n-4)(n-5)\,\phi(n)$$

does so for any (reasonable) function $\phi$. But one's instinctive feeling is that any *naturally occurring* sequence that begins 1, 2, 4, 8, 16, ... will have $\phi(n) = 0$.

Well, one sequence that does start in that way is $(r_n)$. This is obtained by taking $n$ points (not regularly spaced) on a circle and drawing all the $\binom{n}{2}$ lines that join these points in pairs. As long as the original points are sufficiently general, no 3 of these lines will concur at any interior point. The number of regions into which the circle is divided by this procedure will then be the maximum possible and will be denoted by $r_n$.

This gives the sequence $(r_n)$ below. To a person meeting this result for the first time, the fact that $r_6 \neq 32$ is invariably a cause for considerable astonishment and usually leads to furious recounting.

| number of points | $n$ | 1 | 2 | 3 | 4 | 5 | 6 | 7 | 8 | . . . |
|---|---|---|---|---|---|---|---|---|---|---|
| number of regions | $r_n$ | 1 | 2 | 4 | 8 | 16 | 31 | 57 | 99 | . . . |

Of course, having satisfied oneself that the above value of $r_6$ is correct, the problem then is to discover how the sequence $(r_n)$ continues and, if possible, to find a formula for $r_n$. This task will probably keep you entertained for quite a long time.

In case you do not succeed in solving it for yourself—and it is not too easy to find the best approach—it will be only fair to provide you with one of several possible solutions. If you *are* able to conjecture the correct formula for $r_n$ from this table (by constructing difference sequences or otherwise), it is *possible* to prove it correct by mathematical induction. But that does not turn out to be the easiest method in this case. So, despite the title of this chapter, the solution given will be a rather elegant direct method for discovering the formula.

Imagine the figure with $n$ vertices built up line by line. Each line as it is added cuts across a certain number of preexisting regions and thereby increases the count of regions by that same number. It simultaneously creates a number of new interior points of intersection. (Remember that only 2 lines are to meet at any such point.)

Now the increase in the number of regions is always exactly one more than the increase in the number of interior points. For example, in

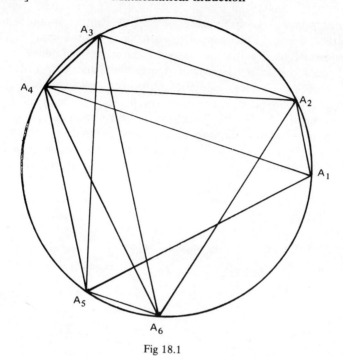

Fig 18.1

the partially completed diagram shown (Fig. 18.1), when the line $A_1A_3$ is added, it will create 2 new interior points and increase the number of regions by 3. If subsequently $A_2A_5$ is added, the number of interior points will then go up by 4 and the number of regions by 5. When a line is added that creates no new interior points (like $A_1A_6$), it will just augment the number of regions by 1.

When the figure is complete, the *total* number of interior points will be $\binom{n}{4}$, since there is a bijection between these interior points and each set of 4 vertices selected from the $n$ given points. (The 6 lines joining any 4 points on a circle always have exactly *one* of their 3 remaining intersections inside the circle.)

Now, the figure starts with just one region and no interior points. Every line adds one new region PLUS one extra region for each new interior point created.

So, when the figure is complete, the number of regions is $r_n$, where

$$r_n = 1 + (\text{total number of lines}) + (\text{total number of interior points})$$

$$= 1 + \binom{n}{2} + \binom{n}{4}$$

This formula gives the observed sequence $(r_n)$. It must be understood, of course, that, if $n < m$, the symbol $\binom{n}{m}$ is to be given the value 0.

It is worth noting that the second difference sequence of $(r_n)$ is actually the sequence $(p_n)$ associated with the pancake problem at the beginning of the chapter. The first difference sequence is $(q_n)$ where $q_n = \binom{n}{1} + \binom{n}{3}$. It is possible to show how $(r_n)$ can be 'built up' from $(q_n)$ and $(q_n)$ from $(p_n)$, thereby establishing the formula for $r_n$ by what is effectively an induction argument. But this alternative demonstration, although interesting, is more cumbersome than the simple proof above.

Euler's theorem provides yet another method of approach.

# Chapter 19

# Numbers and quantities

---

## 19.1

Should a letter in (ordinary) algebra represent a number or a quantity? Does $t = 6$ or does $t = 6$ seconds? Does $g$ stand for 9·8 or for 9·8 m s$^{-2}$? Does the book cost $x$ or does it cost £$x$? This is a dilemma you will have to resolve; yet it is one to which, unfortunately, no pat answer can be given. In several other places, the wisdom of cooperation with your science colleagues has been urged. It is a pity that on this, the most fundamental decision of all, the needs of the subjects are well-nigh irreconcilable.

The problem can often be shelved below fifth form level, although the consequences of this for the pupil are not altogether satisfactory. By boldly confronting the issue, you will come to understand why it generates such passion and may be able to look more sympathetically on colleagues who come to conclusions different from yours.

## 19.2

When a scientist writes any formula, such as

$$\eta = \frac{\pi a^4 p}{8l \mathrm{V}}$$

(Poiseuille's formula) in the theory of viscosity, he is using the letters to denote certain *quantities*: V is the volume of liquid of viscosity $\eta$ flowing per unit time along a tube of radius $a$ and length $l$ when the pressure difference between the ends is $p$. It is understood that some *consistent* system of compound units (nowadays, of course, always SI units) is being used to measure these quantities ($a$ and $l$ in m, V in m$^3$ s$^{-1}$, $p$ in N m$^{-2}$, $\eta$ in N s m$^{-2}$). Each symbol, other than the constants $\pi$ and 8, represents a quantity and a letter is present for *every* quantity that occurs. All the quantities have specific dimensions:

$$[a] = [l] = L, \quad [V] = L^3T^{-1}, \quad [p] = M\,L^{-1}T^{-2}, \quad [\eta] = M\,L^{-1}T^{-1},$$

and the equation is dimensionally correct.

Similarly, when a scientist writes a formula like

$$x = ut + \tfrac{1}{2}ft^2$$

(for motion with constant acceleration), *he* is using $f$ to denote that acceleration and $x$ to mean the distance† travelled in time $t$ when the initial speed is $u$. Again, a literal symbol is present for *every* quantity and the equation is dimensionally correct, each term in this case being of dimension L. According to his canon, the scientist can legitimately write $u = 9$ m s$^{-1}$, $f = -4$ m s$^{-2}$, $t = 2$ s and deduce that $x = 10$ m.

What, however, he *cannot* do is to make a partial substitution for some quantities and not others: to say, for example, that, for free downward motion under gravity, where the constant acceleration is $9.8$ m s$^{-2}$, the above formula becomes

$$x = ut + 4.9\,t^2.$$

Here an illicit attempt has been made to replace $f$ by $9.8$ and the equation is no longer dimensionally correct. The scientist cannot have it both ways; if he has demanded that $f$ be an acceleration, it must remain an acceleration and cannot on the next line become a number. It would, of course, be completely logical for him to replace $f$ by $9.8$ m s$^{-2}$ and write

$$x = ut + (4.9 \text{ m s}^{-2})t^2,$$

but how many zealots do you know who do that?

As long as there is a letter present for *every* quantity, that letter can, so to speak, carry the dimensional burden and everything is fine. The automatic dimension check in the scientists' formulae gives them their great power and elegance. Moreover, in practice, it is usually not too difficult so to arrange a calculation that *all* numeric substitution is done together in one burst at the conclusion of the work, thereby avoiding the sort of solecism just mentioned.

[It is worth pointing out, as an aside, that the value of dimension

†See footnote on page 144.

checks is not confined to formulae in science and applied mathematics. In $\int dx/(x^2 + 9)$, $x$ cannot be taken to be anything but a number, and therefore dimensionless. But, in the general formula $\int dx/(x^2 + a^2)$, it is *possible* to consider $x$ and $a$ as lengths (as one does, in a sense, when the cartesian graph of the integrand is drawn). Since $dx$ may then also be thought of as having dimension 1 in length, it will be seen that $\int dx/(x^2 + a^2)$ is of dimension $L^{-1}$. The correct formula $[1/a \tan^{-1}(x/a)]$ must therefore be of dimension $L^{-1}$: an answer such as $\tan^{-1}(x/a)$ is wrong on dimensional grounds. (This check will not, of course, show up a wrong answer like $1/(2a) \tan^{-1}(x/a)$, where the error is of a different kind.) By contrast, $\int dx/\sqrt{(a^2 - x^2)}$ is dimensionless, as is the formula $\sin^{-1}(x/a)$; this time, $1/a \sin^{-1}(x/a)$ is obviously wrong. For various reasons, however, you may find yourself reluctant to pass on this way of thinking to your pupils (even though it would support your efforts to emphasize that, when $t = \sin^{-1} u$, both $t$ and $u$ must be (dimensionless) numbers—see Section 6.3).]

**19.3**

Returning now to

$$x = ut + \tfrac{1}{2}ft^2,$$

the 'number enthusiast' regards $t$, $x$, $u$ and $f$ all as numbers. The acceleration of *his* particle is not $f$, but $f$ m s$^{-2}$; it travels a distance $x$ m in $t$ s, starting with an initial speed of $u$ m s$^{-1}$. He can put $f = 9.8$ when necessary without any pangs of conscience whatever and when he is given the numbers $u = 9$, $f = -4$, $t = 2$, he calculates that $x = 10$ and *then* says that the distance travelled is 10 m. His handicap is that he cannot write $u = 9$ m s$^{-1}$ and so on, and so has to start most calculations with a blanket statement such as 'All quantities are measured in SI units'.

[Incidentally, to digress once more, you should take the trouble to make yourself thoroughly familiar with SI units, their abbreviations and the conventions for writing them. Notice particularly that a final 's' should NEVER be added to the abbreviation for ANY unit (SI or otherwise) to denote a plural: cm is the abbreviation both for centimetre *and* for centimetres; kg for kilogram *and* for kilograms, and so on. Among several reasons for this are: (1) tabular work looks clumsy if some numbers are followed by 'cm' and others by 'cms'; (2) the SI notation is international and not all languages form their plurals like English by adding an 's'; (3) s is the standard abbreviation for 'second', so that, for example, N s would be a unit of impulse (newton second) and hence must not be used for 'newtons'; similarly, ms means 'millisecond' and cannot mean 'metres'. It is also the recommended practice to adopt this habit with Imperial units (while they still survive)—indeed, with *all* units: if, say, you choose to abbreviate 'year' to 'yr', then make your 'yr' stand both for 'year' *and* for 'years'.

You should note that nowadays k (not K) is used for 'kilo-'. This alteration was illogical, but is now a fait accompli. In the old days, it was always customary to use an upper case prefix for a positive power of 10 and a lower case prefix for a negative power:

$$\text{D deka } 10^1, \quad \text{H hekto } 10^2, \quad \text{K kilo } 10^3;$$

$$\text{d deci } 10^{-1}, \quad \text{c centi } 10^{-2}, \quad \text{m milli } 10^{-3}.$$

This simple convention has been *kept* for all the prefixes in the modern sequence, *except* for kilo:

$$\text{k kilo } 10^3, \quad \text{M mega } 10^6, \quad \text{G giga } 10^9, \quad \text{T tera } 10^{12};$$

$$\text{m milli } 10^{-3}, \quad \mu \text{ micro } 10^{-6}, \quad \text{n nano } 10^{-9}, \quad \text{p pico } 10^{-12},$$

$$\text{f femto } 10^{-15}, \quad \text{a atto } 10^{-18}.$$

The probable reason for this curious change is that in the SI system K was misguidedly chosen as the abbreviation for a kelvin.

And, while on the subject of the prefix 'kilo-', observe that the stress comes on the *first* syllable: kílogram, kílohertz, kílojoule, kílowatt, kílovolt, kílonewton, etc: it should jolly well *stay* there in kílometre!]

### 19.4

The reason why the mathematics teacher is so passionately determined that all his letters shall stand for numbers is that this is the *only* sensible procedure to follow in *elementary* algebra. It is fundamental to our way of presenting the subject at this level that each letter used is representing some number. When quantities are involved, the boy's age is *not* $x$, but $x$ years, so that $x$ can be a number, and so on in all similar circumstances. Let us modify that hackneyed example with apples and bananas to make the point more clearly. If an apple costs $a$ pence and a book costs £$b$, then the cost of 20 apples and 3 books is $(20a + 300b)$ pence or £$(\frac{1}{5}a + 3b)$: one cannot just say 'let an apple cost $a$ and a book cost $b$'.

Unless the ground has been well prepared with simple examples, children will easily become confused when quantities measured in compound units (like density and speed) are involved. One cannot write, as thoughtless pupils are apt to do, 'Let the man walk for a distance $x$ at 5 kilometres per hour: then the time taken is $\frac{1}{5}x$'. For, if $x$ is a distance, then $\frac{1}{5}x$ is another distance; there is no way in which it can become a time. But, if he walks $x$ km at 5 km h$^{-1}$, then the time taken is $\frac{1}{5}x$ h. Here $x$ is a number and everything makes sense.

To drum home this message, consider the following problem.

A man who cycles at 20 km h$^{-1}$ has to make a journey of 9 km. On the way, he has a puncture and pushes the bicycle the last part of the way, walking at 5 km h$^{-1}$. The whole journey takes 45 minutes. How far was he from his destination when he had the puncture?

|  |  |
|---|---|
| *Solution* 1 | *Solution* 2 |

Let the distance walked be $x$ km.

Then the distance cycled was

$$(9 - x)\text{km}.$$

So the time spent walking was

$$\frac{x}{5}\,\text{h}$$

and the time spent cycling was

$$\frac{9 - x}{20}\,\text{h}.$$

Hence

$$\frac{x}{5}\,\text{h} + \frac{9 - x}{20}\,\text{h} = \frac{3}{4}\,\text{h} \qquad (1)$$

$$\Leftrightarrow \qquad \frac{x}{5} + \frac{9 - x}{20} = \frac{3}{4} \qquad (2)$$

$$\Leftrightarrow \qquad 4x + 9 - x = 15$$

$$\Leftrightarrow \qquad 3x = 6$$

$$\Leftrightarrow \qquad x = 2.$$

The puncture happened 2 km
                    from his destination.

Let the distance walked be $x$.

Then the distance cycled was

$$9\,\text{km} - x.$$

So the time spent walking was

$$\frac{x}{5\,\text{km h}^{-1}}$$

and the time spent cycling was

$$\frac{9\,\text{km} - x}{20\,\text{km h}^{-1}}$$

Hence

$$\frac{x}{5\,\text{km h}^{-1}} + \frac{9\,\text{km} - x}{20\,\text{km h}^{-1}} = \frac{3}{4}\,\text{h} \qquad (3)$$

$$\Leftrightarrow \quad 4x + 9\,\text{km} - x = 15\,\text{km} \qquad (4)$$

$$\Leftrightarrow \qquad 3x = 6\,\text{km}$$

$$\Leftrightarrow \qquad x = 2\,\text{km}.$$

The puncture happened 2 km
                    from his destination.

On the left is the normal, recommended Solution 1 with $x$ representing a number. The pattern will be familiar, but notice particularly the statement marked (1). Unfortunately, this line is all too often omitted, yet it lies at the very heart of the solution. The information given in the question says something about the equality of certain *quantities* (in this case, of certain times) and it is from the relation (1) among the *times* that one is able to obtain the equation (2) satisfied by the *number x*.

Can the determined 'quantity fanatic' write down a solution that is consistent with his avowed determination to make $x$ a *distance*? Yes; but as you will see in Solution 2 on the right, maintaining this consistency is an arduous exercise. Note that the distance cycled must be 9 km $- x$, *not* $9 - x$, and the expressions for the times must originate with speeds, not numbers, in their denominators. The relation (3) is again obtained by equating *times*, but now the equation (4) for the *distance x* is obtained by multiplying each side of (3) by 20 km h$^{-1}$ and then simplifying

$$20 \text{ km h}^{-1} \times \tfrac{3}{4} \text{ h}$$

to 15 km.

The writer hopes it is clear that this second method is *not* being seriously offered as a sensible method of solution: it is exhibited to provide a warning of the perils that lie in wait for those who take $x$ as a quantity in elementary algebra and the difficulty of rescuing the solution once that decision has been taken. The real reason why it is not really convenient to *allow* $x$ to be a distance in this problem is that the other quantities (20 km h$^{-1}$, 9 km, etc.) are given numerically and not literally, so the scientists' convention of letting $x$ be a quantity is hopelessly cumbersome.

### 19.5

But even this example is not strong enough to show you how insidious these quicksands are. The seeds of the dilemma are sown long before the beginnings of formal algebra, even back in the primary school. Consider the area of a rectangle and that well-known formula

$$A = l \times b.$$

What is it you imagine this formula is saying? Area equals length times breadth? This indeed is probably how most people regard it, which may show you that, however dedicated a pure mathematician you are, you perhaps sometimes use letters to represent quantities. If you do wish to view the formula in this way, then you will not have to be dismayed to see your pupils write $l = 9$ cm, $b = 7$ cm, $A = 63$ cm$^2$. By doing this, however, they may be creating confusion for themselves later on when other mathematics teachers are going to insist that pupils must use letters only to represent numbers, to prevent the sort of complications discussed above.

The other way to look at the formula, of course, is to regard it as saying that, if a rectangle of length $l$ units and breadth $b$ units has an area of A square units, then

$$A = l \times b,$$

in which case, the teacher will expect the children to answer the previous question by writing $l = 9$, $b = 7$, $A = 63$ and *then* saying that the area is 63 cm$^2$. This is quite unexceptionable to any mathematics teacher. Many will probably insist on following this model as a way of instilling good habits for later algebra, but to other mathematics teachers and most scientists such inflexibility will smack of pedantry.

For the sake of future progress in algebra, it is obviously a good idea to give young pupils early experience of literal representation using simple concrete formulae like this one for area, well before the work is actually called 'algebra'. But even this laudable objective can be seen to have inherent difficulties. However one looks at it, this vexed question cannot be avoided.

Perhaps mathematicians and scientists should long ago all have agreed on some simple convention, such as the consistent use of lower case letters to represent numbers and upper case letters to represent quantities. Then everyone would have been contented, students would have been less confused and the teaching of both mathematics and science would surely have benefited. But global reforms like that stand about as much chance of adoption as a universal language or an international currency!

# Index—contrast or distinction between

# Subject Index